Wasser: Ökologie und Bewirtschaftung

Zunehmend arbeiten Hydrologen, Hydrauliker, Ingenieure des Wasserwesens sowie Hydrochemiker und Biologen in interdisziplinären Gruppen zusammen. Allen gemeinsam ist, dass sie als Ökologen unter Berücksichtigung technisch-ökonomischer Gesichtspunkte die Strukturen und Funktionen der Gewässer erkennen, nutzen, gestalten und erhalten müssen. Die Reihe wendet sich an alle, die sich in Praxis und Theorie mit den Themen Oberflächengewässer, Grundwasser und Wasserversorgung oder Abwasserentsorgung beschäftigen. Das Spektrum umfasst sowohl Konzepte und Anforderungen, die technischer oder politischer Art sein können, als auch Techniken, Methoden und Modelle

Weitere Informationen zu dieser Reihe finden Sie unter
http://www.springer.com/series/14331

Frank Ahlhorn

Integrated Coastal Zone Management

Status, Challenges and Prospects

Springer Vieweg

Frank Ahlhorn
Küste und Raum - Ahlhorn & Meyerdirks GbR
Varel
Germany

The presentation of some equations and structural elements was not correct in the electronic version, this is now corrected. We apologize for any inconvenience and thank the readers for bringing it to notice.

Wasser: Ökologie und Bewirtschaftung
ISBN 978-3-658-17050-9 ISBN 978-3-658-17052-3 (eBook)
DOI 10.1007/978-3-658-17052-3

Library of Congress Control Number: 2017936654

Springer Vieweg

Printed on acid-free paper

This Springer Vieweg imprint is published by Springer Nature
The registered company is Springer Fachmedien Wiesbaden GmbH
The registered company address is: Abraham-Lincoln-Str. 46, 65189 Wiesbaden, Germany

Preface

Intention Imagine you walk along the coast, what do you see? Do you see open space for birds, animals and different types of plants? Do you see meadows and fertile land for crops and "energy plants"? Or do you see a green line at the horizon which protects people against storm surges? What about the ditches, canals and rivers? Are these for draining the land or for leisure ships or the backbone of the northern European water transport sector?

The way we are looking at the coast is strongly depending on our experiences, our awareness, our living and educational environment and many more stakes. The German philosopher Kant already stated that space (here, we understand "the coast" as part of the space of the earth) cannot entirely be grasped or described as "thing itself", only by sensual perception and experiences. Consequently, there is no unique idea or description or perception of the coastal space possible.

Herein we can find two lessons to comprehend:

1. Different living and educational environments lead to different perceptions of the coastal zone. This is a matter of fact a coastal manager has to accept.
2. If we are trying to practice coastal zone management we have to *integrate* the different perceptions, different intentions and ideas in the coastal zone.

If coastal zone management is seen as an instrument for the sustainable development of the coastal zone (UN, 1992, chapter 17) it is only possible as an integrated approach.

This book builds on different classes and lessons I gave at the Carl-von-Ossietzky University Oldenburg (Germany) and the Rijksuniversiteit Groningen (The Netherlands). Students from different countries with a variety of previous knowledge but all interested in coastal issues motivated me to compile this educational book on Integrated Coastal Zone Management (ICZM).

Target Group This book is written for students who are studying in bachelor or master programs related to coastal issues. Although this book is concentrating on water management and not going into detail on different other sectors it provides a methodological part that can be ubiquitously applied. One has to consider the adaptation of the different tools and methods to his or her specific circumstances – as always.

Furthermore, this book intends to provide insights in, generally spoken, coastal water management and their interrelationships for policymakers and managers in coastal zones. Although the spatial focus of this book lies in the Southern North Sea Region, the examples and description should provide relevant information and basic findings on processes (either natural or human) and approaches that could be applied elsewhere.

Last but not least this book may also be read by people who gained long experiences in the coastal zone but would be challenged by a different view at the coast.

Structure This book is divided into eight chapters. It starts with a short description of the context of this book dealing with the coast. What is the coast and how men prepared the area to be able to live and work here. Afterwards, a brief overview about the historical development of ICZM is provided. Sometimes it is difficult to exactly determine a starting point because the idea of an integrated approach starts somewhere and emerges over time at different places. Here, the starting point for the integrated management approach of coasts is set in the developments around the San Francisco Bay in the USA and the first steps taken at the Port Phillip Bay in Australia near Melbourne.

In Chap. 3 basics of ICZM will be explained and discussed as there will be a list of selected definitions of the term ICZM. ICZM will have to be enlivened so basic principles need to be developed and discussed. The reflection will show that these principles are a combination of different relevant processes and approaches for integration.

Chapter 4 summarizes hydrological aspects which are important to know when dealing with the coast. It serves as knowledge base for the following chapters.

In Chap. 5 current challenges in the coastal zone of the Southern North Sea Region will be elaborated. The description is focused on water management, encompassing coastal defense (protection), drainage management and river management. The limitation to the water sector is a tribute to the amount of space of this book. To touch all different sectors and interests in the coastal zone would have been too ambitious. On the one hand, it is striven to elaborate the water chapters in a way that the basic idea of how to treat other sectors will be clear. On the other hand, the water sector is one of the important and basic sectors to enable life and work in the coastal zone (see Sects. 5.1 and 5.4).

Subsequently, existing approaches of integration in the water sector are shown in Chap. 6. Although promising starting points exist there is still a lot to do.

This chapter is followed by the description of a methodological toolbox to enable an integrated management approach (Chap. 7). Those sections should be understood as suggestions and references for the identification of available and adequate methods.

Chapter 8 comprises real-world case studies which could be used to apply the tools and methods touched in the previous chapters.

The book will hopefully stimulate and encourage the reader to further work on and support the integration of different sectors in coastal zone development. Eventually the basic idea of applying or implementing *ICZM* is the aspiration or even achievement of sustainable development of the coastal zone. At the end of each chapter or a respective subsection the exercise questions will motivate to derive a deeper insight of a specific theme.

What This Book Is About and Is Not About! This book will give a brief but as comprehend as possible overview on the item of *ICZM* exemplified for water management in a broader sense. It is neither the ambition to reflect the entire societal nor scientific discussions. It is the ambition to prepare the reader to roughly understand these discussions and hopefully to sort out and assess the different interests and needs behind the oral or written statements of the respective representatives of different sectors. Within this book it is striven for sufficient representation of each point of view. The reference lists provide further literature to deepen the insight of a specific item.

Acknowledgment

The process to write such a text book is impossible without *external* help. First of all I would like to say thank you to my family who has to cope many times when I hide myself away to make progress. Furthermore, I am grateful to my colleagues Leo Adriaanse and Helge Bormann who read first drafts of this book and provided me with helpful and substantial advices. Last but not least, Helge Bormann took over to write the chapter of coastal hydrology, thank you very much.

Varel
Frank Ahlhorn

Contents

List of Figures

Abbreviations

ALARP	As Low As Reasonably Practicable
BCDC	Bay Conservation and Development Commission
CBA	Cost-Benefit Analysis
CPMR/CRPM	Conference of Peripheral Maritime Regions
CZMA	Coastal Zone Management Act
CZME	Coastal Zone Management Effectiveness Study
DAD	Decide-Announce-Defend
DPSIR	Driver, Pressure, State, Impact and Response
EDD	Engage-Deliberate-Decide
EEA	European Environmental Agency
EEZ	Economic Exclusive Zone
EIA	Environmental Impact Assessment
FDA	Dutch Flood Defense Act
FFH	European Flora-Fauna-Habitat Directive
FGG	Flussgebietsgemeinschaft (River Basin District)
Floris	Research project Flood Risks and Safety in The Netherlands
FRMD	European Flood Risk Management Directive
GESAMP	Joint Group of Experts on the Scientific Aspect of Marine Environmental Protection
ICZM	Integrated Coastal Zone Management
IMO	International Maritime Organization
IPCC	Intergovernmental Panel on Climate Change
MCDA	Multi-Criteria Decision Aid
MLS	Multi-Layer Safety Concept of The Netherlands
NAP	Normaal Amsterdam Peil (Amsterdam Ordnance Datum)

NGO	Non-Governmental Organization
NOAA	National Oceanic and Atmospheric Administration
OECD	Organization for Economic Co-operation and Development
PELBDS	Pan-European Biological and Landscape Diversity Strategy
PIA	Participatory Integrated Assessment
PROMETHEE	Preference Ranking Organization Method for Enrichment Evaluation
RCB	Regional Coastal Boards
SEA	European Strategic Environmental Assessment
TAW	Dutch Technical Advisory Commission
TEU	Twenty-Foot equivalent unit
TEV	Total Economic Value
UNCLOS	UN Convention on the Law of the Sea
UNEP	United Nations Environment Program
VCC	Victorian Coastal Council
VNSC	Flanders-Dutch Steering Committee for the Schelde
WBGU	Scientific Advisory Board for Global Change of Germany
WCC	World Coast Conference
WCED	World Commission on Environment and Development
WFD	European Water Framework Directive
WMO	World Meteorological Organization

The Coast 1

1.1 Introduction

Before we start to dive into the complexity of integrated coastal zone management, because the understanding or even the definition of these four words would take us on a long journey, a short introduction of the term *coast* should be given. I don't want to repeat most of the words that have already been written either by poets or scientists, it's only to get a somewhat clearer picture of what is meant when we talk about *the coast* or *the coastal zone*. Lengthy definitions exist for these terms and I'd like to present a selection of them to, afterwards, come to a conclusion on how this understanding might resonate through this book.

The chapter "Encircling the beginning – where (I)CZM started" falls back on two starting points one in Australia and the other one in the USA, at San Francisco Bay. At the end of the first steps according to Integrated Coastal Zone Management (ICZM) in the USA there was the enactment of the Coastal Zone Management Act in 1972. Also in 1972 a workshop was conducted to identify critical problems in the coastal zone – see Ketchum (1972). The participants don't only concentrate on the collection and identification of problems and challenges they also provided a *working definition* for the term coastal zone (Ketchum, 1972, p. 4–5):

> The coastal zone is the band of dry land and adjacent ocean space (water and submerged land) in which land ecology and use directly affect ocean space ecology, and vice versa. The coastal zone is a band of variable width which borders the continents, the inland seas, and the Great Lakes. *Functionally*, it is the broad interface between land and water where production, consumption, and exchange processes occur at high rates of intensity. *Ecologically*, it is an area of dynamic biogeochemical activity but with limited capacity for supporting various forms of human use. *Geographically*, the landward boundary of the coastal zone is necessarily vague. The oceans may affect climate far inland from the sea. Ocean salt penetrates estuaries to various extents, depending largely upon geometry of the estuary and river flow, and the ocean tides may

F. Ahlhorn, *Integrated Coastal Zone Management*,
Wasser: Ökologie und Bewirtschaftung, DOI 10.1007/978-3-658-17052-3_1

extend even farther upstream than the salt penetration. Pollutants added even to the freshwater part of a river ultimately reach the sea after passing through the estuary. The seaward boundary is easier to define scientifically, but it has been the cause of extensive political argument and disagreement. Coastal waters differ chemically from those of the open sea, even in areas where man's impact is minimal. Generally, the coastal water can be identified at least to the edge of the Continental Shelf (depth of about 200 m), but the influence or major rivers may extend many miles beyond this boundary [...].

Within this working definition a distinction was made between functional, ecological and geographical aspects. Here, the vagueness and ambiguity is expressed that a stiff definition and delineation of the coastal zone is neither possible nor practical. If you think of the coastal zone of your country walk along the shore or the beach and try to think about all the interactions and influences both sides of the *edge* have. Some authors facilitate the view "that the boundary lines should be determined by the issues which led to the creation of the [coastal zone management] program. Generally, this means a variable inland line" (Sorensen and McCreary, 1990, p. 6).

For example, Post and Lundin (1996, p. 3) wrote that

the limits of the coastal zone are often arbitrarily defined, differing widely among nations, and are often based on jurisdictional ease. It has often been argued that the coastal zone should include the land area from the watershed to the sea, which theoretically would make sense as this is the zone where biophysical interactions are strongest. For planning purposes this definition is often quite impractical, however, as huge areas containing whole countries would fall under this definition.

Consequently, a definition is hard to determine but the question is: Is it necessary to find or develop a commonly accepted definition for the term *coastal zone* to cover all characteristics and situations?

In the Model Law on Sustainable Management of Coastal Zones (CEC, 2000) the definition of the term coastal zone is in Article 1 as follows:

"For the purpose of this law, coastal zone shall be taken to mean a geographical area covering both the maritime part and the terrestrial part of the shore, including salt-water ponds and wetlands in contact with the sea. [...] The coastal zone shall be precisely delimited at national level. It may extend, according to specific local economic or ecological requirements [...]."

In recent publications on EU level, for instance, one did not seek to provide a definition only a general description is given as

...whole books have been written about the coast and the coastal zone. For our purpose, however, we use coastal zones to refer to the area where the influence of the land and the sea overlap. That would include, for instance, parts of the land affected by sea water seepage and parts of the sea affected by the flow of sand and mud from major rivers" (RIKZ, 1999, p. 7). This statement was made in a publication meant as a preparatory study for "A Coastal Zone Perspective.

Nevertheless, I would like to refer to a recent document published by the European Environment Agency (EEA) where

> ..."the coastal zone [...] is interpreted as the resulting environment from the coexistence of two margins: coastal land defined as the terrestrial edge of continents, and coastal waters defined as the littoral section of shelf areas".

In practice the EEA differentiate on land between a 10 km buffer and an immediate coastal strip of 1 km inland. The area in between these borders is called *coastal inland.*

Later on, it will get much clearer how complicated and difficult a fixed delimitation would be. Either the inland border exceed the area of influence in one way or the whole area up to the alpine region has to be considered. Finally, to conclude we take the minimum description as the coastal zone is the area where land meets the sea. Necessary delimitations and definitions will arise when dealing with all the different issues, problems and challenges at the coast.

For now, we should agree that everybody knows what *the coast* and *the coastal zone* is as long as we don't start to find a pristine definition.

1.2 Characteristics

What are the main characteristics of coastal zones? Here, I would like to touch only four characteristics:

- Gradual transition between land and sea
- Continuously changing environment
- Highly productive area
- Human idea of open space

1.2.1 Gradual Transition

As in the definitions mentioned the coastal zone is at the edge, the meeting place for land and sea. The salt water meets the fresh water coming down the rivers pouring as estuaries into the sea. Where the ocean meets directly the shore a gradual transition can be found at soft coasts such as the Wadden Sea where tidal flats and sand dunes illustrate the transition (Fig. 1.1). In the southern North Sea Region the coast is inundated twice a day by tides. Depending on the energy transported through waves and currents the grain size also displays the transition from sea to land. High energy by waves and currents at the sea meet with slower flowing rivers and increasing tidal flats. At hard rock coasts this kind of transition is either very narrow or absent (Fig. 1.2).

The gradual transition can be observed in chemical and ecological status, where all types of habitats, either saline or fresh or brackish, exist. This transition can easily be experienced in open rivers or ditches which are connected to the sea. This could mainly be found in the estuaries, although they have been altered by human interference.

Fig. 1.1 A gradual transition between land and sea can be found at the Wadden Sea. Tidal flats are slightly transformed by plants to land (salt marshes). The salt marsh is growing under certain conditions and, finally, will not be inundated twice a day by tidal high water, only during storm surges. This type of coast is more vulnerable against erosion and flooding than hard coastal areas like cliffs. © Frank Ahlhorn

Fig. 1.2 The cliffs of the Normandy (France) show a different picture of the coast. The cliffs provide a clear border between land and sea. © Frank Ahlhorn

1.2.2 Continuously Changing Environment

One of the consequences that the high energy of waves and currents meet the land is change in shape and structure of the coast. Beaches show a different appearance in winter or in summer. In winter, storms rework the material at the shore, the higher energy then leads to coarser material on the shore and the beach is much more steep. In summer, long gentle beaches with finer grained material can be experienced. This is the way it may look like at soft coasts such as the Wadden Sea. Also, sand and mud plates are highly mobile. Depending on waves and currents they change in a short-time period shape and structure.

These short-time changes are superimposed by long-term morphological changes. Additionally, humans interfere in past centuries with the morphological development of coasts.

In accordance with the abiotic changes the ecological change occur. Animals and plants adapted or have to adapt to this changing environment, for example, the semi-diurnal change of salt concentration due to the tides or wet and dry conditions.

1.2.3 Highly Productive Area

The connection to nutrient-rich systems like the ocean and the estuaries coasts benefit from that input. For example, the biological productivity of the Wadden Sea is estimated as high as in rain forests. Millions of migratory birds visit on their flyways the Wadden Sea to rest and feed.

On the other hand, former salt marshes were reclaimed first to enlarge the area of influence of an emperor and second to gain highly fertile soil. Today, the highest fertile soils are found on the marsh lands at the coast.

1.2.4 Open Space

Since mankind settled at the sea an impression of the wild and free ocean accompanies that feeling. It is an ambivalent feeling dealing with endless possibilities and the fear of the power of the ocean in times of storm tides. Thorough descriptions can, for example, be found in Michelet (1861/2006).

1.3 Preconditions for Living and Working at the Coast

When we think of the coast we imagine the current shape and structure as given. Over a long time men have rebuilt the natural shape of the coastal landscape (Fig. 1.3) of the Wadden Sea. The natural transition zone between sea and land was built through the tidal flats and the salt marshes. This gradual transition was supported and self-sustained by sediment transport through tides, waves and currents. Tidal creeks gravitationally drain the hinterland.

Many centuries ago men settled in the low-lying coastal area and had to cope with inhospitable circumstances such as tides and storm tides in winter time and the shortage of fresh water.[1] Nevertheless, starting with the first settlers the colonization of the coast began. Commonly supposed the first settlers were peasants either growing sheep or cattle or excavating saline soil for selnering.[2] What about the basic pre-conditions that a colonization took place and was successful?

[1] Find a brief description on p. 79 and in respective literature.

[2] See p. 80.

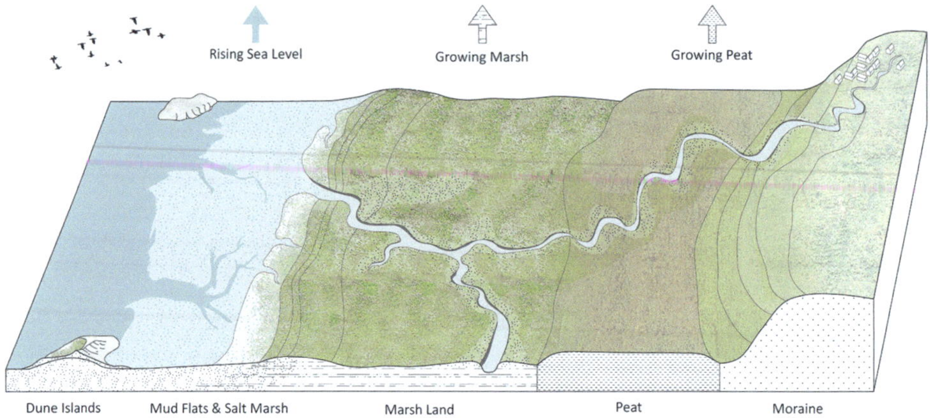

Fig. 1.3 The cross section of the Wadden Sea coast without any protective elements such as sea walls or dikes. From *left* to *right*: the Wadden Sea as transition zone between the open sea and the land consists of sand and mud flats intersected by tidal creeks. Salt marshes build the first manifestation of dry land, although the lower and the middle part of the salt marshes are regularly flooded by tides. Tidal creeks penetrate undisturbed into the hinterland gravitationally draining the moraine and peat areas. The morphological development of the Wadden Sea is driven by the tides, waves and the currents. Natural dynamics of the transition zone would be that tidal flats and salt marshes grow up with the sediment transport by, for example, an increasing sea level. Additionally, with an increasing sea level the entire system of tidal flats, salt marshes and marshes will move land inward. Source: adapted from Reise (2015). © Urbane Landschaften

1.3.1 Coastal Protection

The first settlers had to fight against high tides as they were trying to build farm houses in the low-lying coastal area. Animals sheltered on higher grounds against high tides, also settlers built their houses on higher ground. This ground was heightened continuously trying to cope with a rising sea level or with the water level of severe storm tides. Subsequently, the area of influence (grazing area) should be enlarged and an increasing amount of cattles has to be protected against high water levels. First dikes were built to protect a greater area against flooding. This development (although not smooth[3]) went on and today there is a closed line of main dikes protecting the hinterland.

Thus, *coastal protection (defense)* was one of the first interests and needs people started with at low-lying coastal areas to protect themselves against high water levels from the sea. It was a necessary consequence of settlement and the enlargement of the growing economic wealth in low-lying coastal areas. The protection against storm tides is based on a single line of defense (main dike) which is able to withstand a certain amount of physical load imposed by tides, waves and currents. The successful effort of protecting people

[3]See p. 63.

and values against damage generates new and innovative types of land use in addition to existing ones such as agriculture, trade and shipping. But the closure of the single dike line at the coast demands for innovation in water management, especially in drainage management (see Chap. 5).

1.3.2 Drainage Management

The consequence of closing the dike line demands for innovation in drainage management. Prior to the closure most ditches and rivers were able to discharge gravitationally into the sea (Fig. 1.3). The land was higher than the water level. After closing the dike line wooden tubes and wooden tidal gates were built in the dike line to discharge the inland water. Artificial installations were necessary to drain the hinterland. The area behind the dike was not further supplied by sediment and was drained; thus, the soil surface began to subside. At a certain point the elevation of the land was lower than the water level outside of the drainage means. A further innovation was demanded; the wind mill was installed to actively pump the water from low-lying areas into higher ditches and canals. Consequently, high efforts in draining demand even higher efforts in draining because the land dropped continuously (and sea level rises). Today, large areas of the southern North Sea coast would only be inhabitable with major restrictions.

Thus, *drainage management* is the second necessary interests and needs for people living and working in low-lying coastal areas (Fig. 1.4). The term *water management* is used ambiguously: sometimes it encompasses all aspects of water management such as surface, sea and ground water. Sometimes it touches only the quantity (mainly too much water, e.g. risk at flooding) but also the quality of water (pollution by toxic or fertilizing substances). Sometimes it contains besides the draining of low-lying areas also coastal protection (i.e. the protection against flooding from the sea with all its aspects, e.g. building dikes, fighting erosion) and, additionally, the irrigation of arid areas. Nevertheless, the management of the water is essential at the coast to enable safe living and working. That's why this textbook on ICZM concentrates mainly on water-related issues, but does not lose sight of other interests and needs at the coast.

1.4 Different Types of Land Use

Based on the previous description further interests and needs were established at the coast with the growing safety against flooding and the supply guarantee of essentials of life such as fresh water and food. Hereafter, only a selection of interests and needs is briefly described. The intention is to outline a short overview of aspects which should be kept in mind when talking about the management in the water sector. Some of these land uses are mutually dependent such as agriculture and drainage management. Some of these land uses

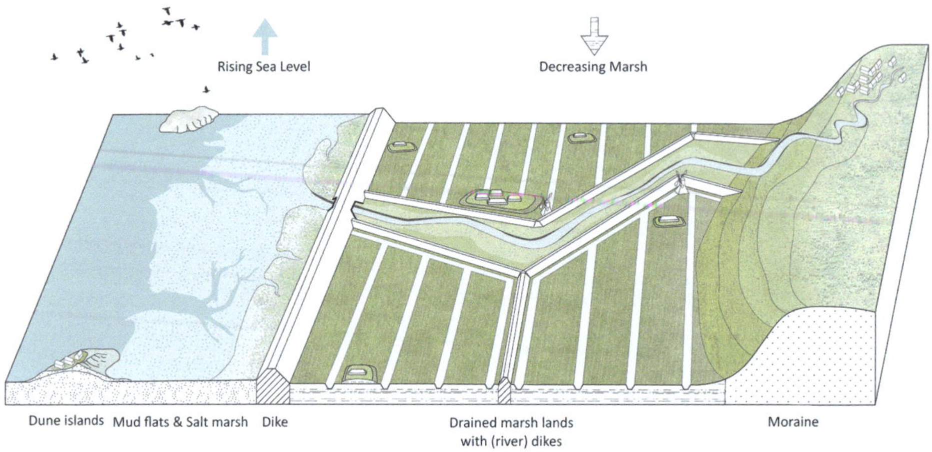

Fig. 1.4 The cross section of the Wadden Sea coast with protective elements such as dikes. From left to right: the Wadden Sea as transition zone between the open sea and the land consists of sand and mud flats intersected by tidal creeks. An artificial border, that is, main dike, disrupts the free movement of water from both sea and land. Consequently, water management means have been installed to control the water level from sea (coastal protection) and from the hinterland (drainage management). Tidal creeks are straightened, narrowed and encircled by river dikes. Marsh lands are disrupted from the sediment supply by tides and drained for different types of land use. First settlers built dwelling mounds (small hills) as protection against flooding. Wind mills are one of the first innovations to artificially drain the hinterland because soil surface is below the water level in the drainage canals. Source: adapted from Reise (2015). © Urbane Landschaften

are not commonly seen as interests and needs such as nature conservation. Nevertheless, linked to all of them is the interest of human kind to use the services and functions the coastal ecology provides.

1.4.1 Agriculture

One of the basic and first interests and needs in coastal areas was agriculture. First settlers started to grow cattle and sheep and excavated saline soil to harvest salt. Today, coastal areas provide high fertile soil for agricultural use. For example, the marsh lands of Lower Saxony are regarded as fertile as the loess soils near Magdeburg further south in Germany.

Typical for coastal farming was the growing of cattle, sheep and dairy farming. Later on, with increasing ability of technical innovations the heavy soil of the marsh lands were used as well for crops. Today, a mixture of dairy and crop can be found strongly affected, for example, in Germany, by the energy transition from fossils to renewable. Meadows are plowed to grow corn which can be used in biomass power plants. Essential for the first settlers and for dairy farming was dry land and a sufficient supply with fresh water. On the one hand, crop farming needs dry land, but in certain times fresh water for irrigation is needed. On the other hand, many areas are used today as mono-culture to cultivate energy plants which affects the hydrological cycle, for example, by increasing the run-off and decreasing the ground water recharge.

1.4.2 Trade and Shipping

Since centuries trade and shipping is closely interwoven and the coast plays a major role in the globalization of trade. Since the Hanseatic League from the middle of the twelfth to middle of the seventeenth century dominated sea trade it is continuously growing. Small sailing boats built the connection between first harbors along the coast. These boats had a limited draft and were able to sail through the dangerous and highly mobile tidal flats. With the growing trade between different countries the harbors were growing. Consequently, the sailing boats increased to carry more goods. The development did not stop until today; container vessels replacing the sailing boats are constantly getting bigger and bigger. Today, the largest container vessels are able to carry about 19,000 container TEU (twenty-foot equivalent unit) with a length of approximately 400 m and a width of approximately 60 m. Another consequence is that the draft of these ships was increasing, for example, from 12 m in 1970 up to 16 m in 2015.

The development of the ship size demands for a higher effort in the provision of safety and easiness of shipping. Thus, the accessibility of harbors needs continuously to be improved, that is, deepening of fairways. Some world leading harbors, especially in the southern North Sea Region, are located more than 100 km inland, for example, Hamburg or Antwerp. Therefore, the river channels are deepened and straightened to enable today's container vessels to safely navigate through the estuaries of the rivers Elbe and the Schelde. This of course has an influence on the hydro-morphological and ecological functioning of the estuaries.

1.4.3 Industry

The increasing economic wealth was accompanied by economic diversification and vice versa, and goods were not solely manually produced but also by industrial production. Special industrial branches are located at rivers or the sea because of the connection to international trade. Some of them need the connection because of the size of their products, for example, aircraft industries to transport large fuselage shells. Some industrial branches are dependent on cooling water which could be abstracted from the rivers. Other branches are located at rivers because the amount of their raw material is best delivered by bulk carrier.

1.4.4 Nature Conservation

Nature conservation is also subsumed as interests and needs at the coast. In ancient times nature conservation was not known, but with the growing understanding of natural processes and that humans are depending on natural provisions some people started to preserve rivers, landscape and flora and fauna. On the one hand, nature conservation is conducted by different organizations called non-governmental organizations (NGO) such as Friends of the Earth or World Wide Fund for Nature. On the other hand, nature conservation is since decades a task of governments, for example, in Germany the national law on nature conservation was first introduced in 1935 and got finally into force in 1977. In Germany the Federal States developed own regulations for nature conservation which are equivalent to the national law in the light of the concurrent legislation.

The five coastal Federal States in Germany have their own nature conservation laws touching the conservation and protection of their coastal areas, for example, the law on the national park in Schleswig-Holstein or Lower Saxony. Also on land several rules and laws exist which protect specific species and landscapes against deterioration. That nature conservation in the Wadden Sea depends upon the water is obviously clear, but also in the hinterland some biotopes and habitats are depending on specific water conditions, for example, wet shallow areas on meadows for wading birds.

Conflicts between nature conservation and coastal defense are mainly upon the way of protecting people against flooding. For example, the closure of a bay would transform a saline water environment into a fresh or brackish water environment. On the one hand, the benefit for coastal protection would be that the main dike line will be shortened which might save money for maintenance. On the other hand, a dramatic change for the saline water habitat would take place with severe consequences for endemic plants and animals (see Sect. 8.4 or the consequences caused by the implementation of the Delta Plan in The Netherlands since 1955). Furthermore, salt marshes are called dike fore land by coastal engineers as they are important to support the main dike in protecting people against flooding (e.g. energy dissipation of waves before attacking the outer slope of the dike). By mutual agreement a broad salt marsh (fore land) is desirable, but with conflicting ways of achievement. For coastal protection the solidification of the edge of salt marsh would be preferable to counteract erosion, where nature conservation would like to support natural dynamics.

1.4.5 Tourism and Recreation

The first recreational use of the Wadden Sea coast dates back to the eighteenth century. The islands were discovered by the upper class to serve as a spa due to the saline environment. Also today the Wadden Sea islands are attractive locations for tourists and recreational use. The main land coast also got more attractive for tourists since the harmful diseases were repelled. For a certain amount of time the coast was seen as an inhospitable area suffering shortages and inhabited by the poor.

Today, tourism and recreation are growing economic branches at the coast. The need for building areas and for fresh water is increasing. Consequently, increasing values for tourist infrastructure demand for the protection against flooding either by the sea or by rivers.

1.4.6 Final Remark

These are only some interests and needs described to provide an overview of the interdependencies between different types of land use. Only some interlinkages have been touched upon. In the respective exercises of each section or subsection these aspects will be touched upon again. Especially, in Chap. 8 the fictitious case studies will elaborate more on these relationships.

References

[Council of Europe Committee of Ministers] CEC. Model Law on Sustainable Management of Coastal Zones and European Code of Conduct for Coastal Zones. *Nature and Environment*, 101, 2000.

B.H. Ketchum. *The water's edge: critical problems of the coastal zone*. MIT Press, 1972. ISBN 0262110482.

J. Michelet. *Das Meer*. Campus, 1861/2006. ISBN 978-3-593-38132-9.

J.C. Post and C.G. Lundin. Guidelines for Integrated Coastal Zone Management. ESD Studies and Monographs, No. 9, The World Bank, 1996.

K. Reise. *Kurswechsel Küste*. Wachholtz-Verlag, 2015. ISBN 978-3-529-05394-8.

[Rijksinstituut voor Kust en Zee] RIKZ. A Coastal Zone Perspective. Preparatory Study. Technical report, RIKZ, 1999.

J.C. Sorensen and S.T. McCreary. *Institutional Arrangements for Managing Coastal Resources and Environments*. National Park Service, U.S. Department of Interior, 1990. ISBN 0-931531-00-4.

Historical Annotations 2

2.1 Encircling the Beginning: Where (I)CZM Started

Despite the fact that the use and to some extent the "management" of the coastal zone started long time ago it is not easy to set a starting point for the beginning of integrated coastal zone management. For example, in the late nineteenth century in Northern Germany civil engineers started to draw plans for the straightening and deepening of river channels – see Franzius (1888/1991). Many years before, men started to build dikes and to reclaim land from the sea and rivers to enhance the territory of emperors and simultaneously the fertile area to be used by peasants (Kramer and Rohde, 1992; van de Ven, 1993).

In these times men admire the capability and ability of human kind to conquer nature. Principally, only the benefits of these new developments were seen and acknowledged. Although, in the late nineteenth century critical voices exist that pinpoint on controversial effects – see Blackbourn (2006). One can also find evidence for critical voices against the typical proceeding of protecting sandy islands in Northern Germany by the implementation of hard measurements, for example, Bartels (1881). Especially, the notion of technical dominion of natural processes was questioned. This aimed at trying to take further aspects into account. At any time the reason to get active was to adapt or develop the natural environment to the purpose of human kind. If negative effects did occur or were detected they were mainly answered by sectoral measurements (see Chap. 3).

2.1.1 San Francisco Bay (USA)

The earliest efforts in *ICZM* dated back to the 1960s in the USA and Australia (Sorensen, 1993). The San Francisco Bay suffered increasing siltation. This led to a reduction of the size of the bay by one-third and destroyed approximately 90 % of tidal marshes. In 1961 a

F. Ahlhorn, *Integrated Coastal Zone Management*,
Wasser: Ökologie und Bewirtschaftung, DOI 10.1007/978-3-658-17052-3_2

plan was generated to fill 60 % of the remaining Bay and create a narrow shipping channel until the year 2020. At that time three women started a campaign against these plans and in 1965 the Bay Conservation and Development Commission (BCDC) was established which was charged to prepare a plan for long-term use of the San Francisco Bay.

A variety of factors led to the decline of the Bay area and the degradation of tidal marshes such as building upstream dams and diversions (Mount, 1995). Further impacts were results of hydraulic gold mining which showed negative effects on anadromous fish populations or of industrial and agricultural activities such as the disposal of trace elements or pesticides.

The San Francisco Bay Plan was completed in 1968 and sent to the California Legislative and the Governor in 1969. After finalizing the studies and, there upon, the Bay Plan the BCDC was designated as responsible agency for maintaining and carrying out the provisions of the law and the Bay Plan.

The following compilation on the San Francisco Bay Plan is based on the report "San Francisco Bay Plan by the San Francisco Conservation and Development Commission" (BCDC) reprinted in 2012 (BCDC, 1969). Here, a short description of the important items of the Bay Plan will be given; further and detailed information can be found at specific websites.[1] Due to the fact that the Bay Plan was developed on the (urgent) issue of Bay filling the major conclusion and policies are touching these points (see BCDC, 1969):

The Bay The Bay is a single body of water, and a Bay Plan can be effectively carried out only on a regional basis.

Uses of the Bay The most important uses of the Bay are those providing substantial public benefits and treating the Bay as a body of water, not as real estate.

Uses of the Shoreline All desirable higher priority uses of the Bay and shoreline can be fully accommodated without substantial Bay filling and without loss of large natural resource areas. But shoreline areas suitable for priority uses – ports, water-related industry, airports, wildlife refuges and water-related recreation – exist only in limited amount and should be reserved for these purposes.

To enable economic and societal development around the Bay one conclusion touches the point of justifying Bay fill: "Some Bay filling may be justified for purposes providing substantial public benefits if these same benefits could not be achieved equally well enough without filling." The explanation of justifiable Bay fills encompass six items such as ports, routes, airports and recreational features. Thereafter, the *effects of filling the Bay* regarding the previously mentioned items are described. So, these items have to be considered as minimum requirements in the application of planned Bay filling. The Commission already acknowledged that the *pressure to fill* the Bay might increase over time. However, the main conclusion concerning the pressure is that the Bay is vulnerable to filling and diking because of the historical development. Approximately 40 % of the Bay was lost due to

[1] For example, www.bcdc.ca.gov

land reclamation and filling of former tidal areas and marshes. If it was decided to fill on areas for a certain reason the Bay Plan asks for safety measurements to secure people and the infrastructure against various hazards. Although the Bay Plan is not directly dealing with the reasons and effects of water pollution either by point or by diffuse sources, this point is addressed by making links to existing regulations and responsible organizations and administrative bodies.

Acknowledging the various triggers on the situation in the San Francisco Bay and the consequences if envisaged developments will not happen the BCDC consists of a variety of interest groups and representatives of different governmental and administrative levels. The BCDC has jurisdiction over a certain area that does not only include the water body of the Bay, but also a 100 feet buffer landward the mean high tide line (for an exact definition, see BCDC 1969). The BCDC and the Bay Plan are still working.

On national US level, starting in the late 1950s, the view of politicians and researchers turned from land to sea side. The ocean got more and more into the focus: *We are just at the threshold of our knowledge of the oceans ... (This) knowledge is more than a matter of curiosity. Our very survival may hinge upon it. (Letter to the President of the Senate and the Speaker of the House of Representatives in 1961 cited in Merrell et al. 2001)*

This statement was an important stimulation to raise the funding for ocean and marine investigations. Several years later, in 1966, the Act on the marine resources and engineering development was signed by President Johnson and came into force. Based on this Act the Commission on Marine Science, Engineering and Resources was established. This Commission, known as "Stratton Commission," released a report on *Our Nation and the Sea* "that was a comprehensive, forward-looking report that reviewed the status of most areas of American Ocean policy and offered 126 recommendations ..." (Merrell et al., 2001, p. 14). Besides the recommendation to establish a national agency for the administration of the ocean and atmosphere (known as NOAA) the Stratton Commission recommend to develop a national *Coastal Zone Management (CZM)* program. Finally, in 1972 the USA enacted the *Coastal Zone Management Act (CZMA)*.

(The) Congress declared four basic national Coastal Zone Management policies in the CZMA (Archer and Knecht, 1987, pp. 104–105):

- To preserve, protect, develop, and where possible, to restore or enhance the resources of the coastal zone of the United States
- To encourage and assist the states to develop and to implement CZM programs meeting specified national standards
- To encourage the preparation of "special area management plans" to protect nationally significant natural resources, to ensure "reasonable coastal-dependent economic growth" and to provide "improved protection of life and property in hazardous areas and improved predictability in governmental decision-making"
- To encourage the participation and the cooperation of public, state and local governments, interstate and other regional agencies, and federal agencies in achieving the purpose of the CZMA.

The core point of the CZMA is that the states are recognized as the *key player* for the development and implementation of CZM programs. Through the CZMA the federal government provides funding for the development of state CZM programs. By the federal consistency doctrine it should be assured that federally permitted activities are consistent with the approved state CZM programs.

Based on the CZMA a federal office was established that oversaw the development and implementation of state CZM programs. According to the CZMA the federal office was responsible for the supervision of the federal funds, to provide support for the development of state CZM programs and to ensure that the states administer their programs as approved. The state CZM programs have to be reviewed and approved by the secretary of the federal agency and, furthermore, to stimulate innovation states could use a certain amount of federal grants for significant improvements as determined by the federal office (Archer and Knecht, 1987).

2.1.2 Port Phillip Bay (Australia)

More or less at the same time on the other side of the earth, in Australia, the Port Phillip Bay Authority started an investigation on the Bay development in 1966. At the Port Phillip Bay the city of Melbourne with approximately 3.7 million inhabitants is located. The catchment area is of approx. 10,000 km^2 (VCC, 2014). The main trigger to conduct the first environmental study between 1968 and 1971 was the reduction of pollution problems either by diffuse or by point sources (NRE, 2002). On the other hand, the shorelines around Port Phillip Bay experienced erosional trends. Alongside the Bay recreational facilities were allowed to be established and, though, ought to be protected by hard structures such as groynes and sea walls (Gourlay, 1996). In the middle of the 1970s the protection of the foreshore was complemented by sand nourishment. Nevertheless, one of the major problems was the pollution of the Bay by nitrogen and contaminant load from the sewage treatment plants in the catchment area. The upgrade of the sewage treatment plants and the approach of *continuous improvement* of processes in the Australian economy showed success and improved the ecological status of the Bay.

This process held until the 1990s when the second comprehensive study was started – the Port Phillip Bay Environmental Study. It was conducted by the CSIRO and was finished in 1996.[2] The results and recommendations of this study became a relevant part of the ongoing management of the Port Phillip Bay (see NRE, 2002). Consequentially, the following environmental risks have been identified for the Port Phillip Bay: deterioration of waters, presence of litter, exotic marine organism, physical disturbance of habitats and harvesting activities.

The Port Phillip Bay Authority, established in 1966, consisted of representatives of government agencies and community groups (Gourlay, 1996). The Bay Authority "had

[2]csiropedia.csiro.au/port-phillip-bay-environmental-study/

the responsibility for coordinating development on public land 600 m to seaward and 200 m inland" (Gourlay, 1996, pp. 20–21). Until the 1980s administration and planning was scattered to many agencies and also on different levels. With the publication of the Victorian Coastal Management Act in 1995 two levels of agencies were established to implement the objectives of the Act: the *Victorian Coastal Council* (VCC) and the *Regional Coastal Boards* (RCB).

The VCC is the "statewide lead agency ... whose primary role is to prepare for the Government the major strategic planning document for Victoria: the *Victorian Coastal Strategy*" (Wescott, 2004, p. 97). The RCBs consist of representatives from local governments and three members of the RCB are sitting in the VCC to ensure a strong link between state and local level. "The RCBs role is to implement the Coastal Strategy regionally and to ensure local input to decision-making is received" (Wescott, 2004, p. 97).

An important document is the Victorian Coastal Strategy (VCS, recently 2014) which should bridge the international, national and regional (i.e., Victorian) approaches of coastal planning and management. In the VCS 2014 it is said that "ICZM is the basis for coastal planning and management in Victoria and is achieved through formal and informal collaboration and coordination between all different groups who use and manage the coast" (VCC, 2014, p. 11).

Further information on Victorian ICZM approach and the Port Phillip Bay can be found at specific websites.[3] A comprehensive description of coastal management in Australia can be found in Harvey and Caton (2010).

Exercise

The rough descriptions of the first steps in (I)CZM in the USA (San Francisco Bay) and in Australia (Port Phillip Bay) show different reasons (sedimentation, erosion and water quality problems) and different triggers (bottom-up and top-down[4]) to start such a process. Given the provided sources of information and the available information on respective websites in the Internet prepare a comparison of both approaches.

- What are the differences and similarities of both ICZM approaches?
- What about the current status of the Bays?
- What are the shortcomings and barriers for the implementation of these ICZM approaches?
- What are the lessons learned based on the result of this comparison?

[3]www2.delwp.vic.gov.au/, ccb.vic.gov.au, www.vcc.vic.gov.au

[4]*Bottom-up* is understood as initiative by stakeholders with no legal status, for example, private persons; *top-down* is understood as action initiated by, for example, governmental bodies or from a central point upper in a hierarchy.

2.2 Brief Outline of International ICZM Steps

Since the first steps were taken, many initiatives and efforts have been started over the world – see, for example, OECD (1993) and Sorensen (2000). Significant events happened and several relevant international and national agreements were signed for the proliferation of ICZM since then. The following listing is an excerpt of what happened according to ICZM worldwide; further information on international conventions is listed in, for example, FAO (2006):

1948/1958 UN Convention in Geneva for the adoption of the International Maritime Organization (IMO), entry into force in 1958.

1965 Establishment of the San Francisco Bay Conservation and Development Commission to conduct an integrated management approach for the protection of the Bay.

1966 Establishment of the Port Phillip Bay Authority for the coordination of developments on a certain strip around the Port Phillip Bay.

1968 Stratton Commission in the USA released the report *Our Nation and the Sea* containing the recommendation for the initiation of a national coastal zone management program.

1969 Bonn Agreement for cooperation in dealing with pollution of the North Sea by oil (replaced in 1983 by a new agreement covering also other harmful substances).

1972 US Coastal Zone Management Act was enacted.

1972 Oslo Convention adopted in 1972 was to control the dumping of harmful substances from ships and aircraft into the sea.

1972 United Nations Conference on the Human Environment resulting in the "Stockholm Declaration" containing 26 principles concerning the environment and development.

1972 RAMSAR Convention on wetlands of international importance, especially waterfowl habitat.

1973 MARPOL is the international Convention for the prevention of pollution from and was modified by the protocol of 1978.

1973 Resolution on the protection of coastal areas by the Council of Europe.

1974 Paris Convention on land-based sources of marine pollution.

1974 Helsinki Convention on the protection of the marine environment of the Baltic Area (renewed and signed in 1992).

1976 Barcelona Convention for protection of the Mediterranean Sea against pollution.

1982 United Nations Convention on the Law of the Sea (UNCLOS) defines the rights and responsibilities of nations with respect to their use of the world's oceans.

1984 First Conference on ICZM in South and Central America.

1985 Workshop in the USA to define the term ICZM.

1987 Publication of the report *Our Common Future*, known as *Brundtland Report* (WCED, 1987), from the UN World Commission on Environment and Development.

1988 Establishment of the Intergovernmental Panel on Climate Change (IPCC) by the UN World Meteorological Organization (WMO) and the UN Environment Program (UNEP).

1990 Publication of the First Assessment Report (FAR) of the IPCC.

1992 UN Conference on Environment and Development (UNCED), known as Earth Summit Rio 1992, where the *Agenda 21* and the Rio Declaration on Environment and Development were adopted. The Agenda 21 contains different chapters with recommendations and statements to several issues, especially Chap.17 is relevant for ICZM. It provides statements according to the protection of the oceans, all kinds of sea, including enclosed and semi-enclosed areas, and coastal areas and the protection, rational use and development of their living resources:

> … International law, as reflected in the provisions of the United Nations Convention on the Law of the Sea referred to in this chapter of Agenda 21, sets forth rights and obligations of the States and provides the International basis upon which to pursue the protection and sustainable development of the marine and coastal environment and its resources. This requires new approaches to marine and coastal area management and development at the national, subregional, regional and global levels, approaches that are integrated in content and are precautionary and anticipatory in ambit, as reflected in the following program areas: (a) Integrated management and sustainable development of coastal areas, including exclusive economic zones … (e) Addressing critical uncertainties for the management of the marine environment and climate change, (f) Strengthening international, including regional, cooperation and coordination …. (UN, 1992, Chap. 17, p. 238)

1992 Adoption and ratification of the OSPAR Convention which is the combination and the update of the Oslo and Paris Convention.

1993 On the *World Coast Conference (WCC)* in "Noordwijk (The Netherlands) the "Noordwijk Guidelines on Integrated Coastal Zone Management" were developed and adopted by participants from 90 coastal nations, 20 international organizations and 23 NGOs (see IPCC, 1994).

1994 The European Union requested the Member States in the Council Resolution of 6 May 1994 on a Community strategy for integrated coastal zone management (94/C135/02) (see also EC, 1995).

1996 The European Commission initiated a Demonstration Program on Integrated Coastal Zone Management from 1997 till 1999 (EC, 1999).

2000 Getting into force of the European Water Framework Directive (WFD) (EC, 2000).

2002 The recommendations on ICZM (2002/413/EC) were approved by the European Parliament and the Council and published in June (EC, 2002).

2007 Getting into force of the European Flood Risk Management Directive (FRMD) (EC, 2007b).

2013 The European Commission released a proposal on the establishment of a framework for maritime spatial planning and integrated coastal management (EC, 2013).

Exercise

- Categorize them according to *binding* and *non-binding* documents.
- Find out which instruments exist if signatories do not obey the content of these documents.

2.3 European Developments

Certainly the occupation with coastal issues started on European level also in the 1960s or earlier. The first visible document is the *Resolution on the protection of coastal areas* by the Council of Europe in 1973 (CEC, 1973). Therein, it was concluded that "the situation is liable to deteriorate still further in the future, having regard on the one hand to the scarcity of coastal areas and the vulnerability of the concentration of human activities on those areas". Furthermore, "the protection of the coast can only be effective if multiple interests and problems are taken simultaneously into account (maintenance of the ecological and biological balance, preservation of the beauty of landscapes and conservation of natural resources, promotion of economic and tourist development, safeguarding of the hinterland)" (CEC, 1973, p. 96).

Based on these conclusions the resolution contains 16 recommendations on how the European Member States should orientate their coastal policies.[5] Here, only a few are highlighted:

- Institute appropriate machinery to co-ordinate the various actions concerning the coastline whether they are initiated by the State or local authorities
- Regulate development in coastal areas:
  - by issuing development bans applying to appropriate strips of land along the seafront and
  - by subjecting the granting of development permits to particularly stringent conditions
- Review systematically the uses to which public land in coastal belts is put, in order to further the policy of protecting and improving those areas
- Adopt special measures to protect coasts from erosion and landslides:

[5]The entire list of these recommendations can be found in specific websites; see, for example, http://ec.europa.eu/environment/iczm/background.htm

  - by stabilizing sand dunes,
  - by regulating excavation and the removal of sand,
  - by prohibiting the cutting and uprooting of vegetation
- Undertake a vigorous campaign to inform and stir public opinion in regard to the protection of the seafront and encourage all public and private initiatives to safeguard the coastline, especially in the form of the creation and management of protected areas
- Co-operate closely with one another where their coastal areas adjoin, with a view to:
  - harmonizing their various sets of regulations and co-ordinating action with regard to the protection of sites, flora and fauna and to pollution control,
  - undertaking where appropriate joint action such as the management of international parks or the pooling of supervisory and pollution control services

The excerpt of these recommendations shows that the resolution from 1973 contains also far-reaching and cross-sectoral approaches for the protection and management of coastal areas in Europe. Some of the Member States already established sectoral administrative bodies and regulations, but the resolution tried to motivate to go beyond the activities of that time.

Follow-ups of this resolution were European Community action programs which focus on the planning and management of coastal areas. These action programs led to the *European Charter of the Coast* in 1981 (CRPM, 1981). The Charter contains ten objectives for coastal zone development in the European Union. Subsequently, in 1982 a *Resolution on the European Coastal Charter* was adopted by the European Parliament (CEC, 1982). In this resolution it is stated that the responsibility for the implementation of the notion of the Charter is with the national, regional and local governments, but that the European level should support these actions.

The resolution indicates three fields where EU institutions should provide support: knowledge, planning and control. In the field of knowledge they promoted the co-operation and exchange between different research centers and initiated an information campaign. In the field of planning the European Parliament encouraged to prepare a "Community program for the integrated development of coastal regions taking into account the need to restore the balance between the hinterland and the coast ..." (CEC, 1982, p. 125). Furthermore, attention should be put on the protection of people and infrastructure in coastal zones which were at risk of flooding. And, for example, the European institutions should provide necessary coordination and finance for the pilot operation mentioned in the Coastal Charter. Regarding control the institutions should devote on a state-of-the-art investigation on existing rules and regulations for coastal areas in the Member States, to improve navigational safety and to reduce land-based pollution of coastal zones. Also the motivation of public participation in coastal planning was touched.

Between the late 1970s and the beginning of the 1990s the European Union established several Directives which touch the coastal zone and show specific focal points – see, for example, Wild Birds Directive from 1979 (oldest European Directive 79/409/EEC current codified version 2009/147/EC), the Flora-Fauna Habitat Directive from 1992 (92/43/EEC) or on Environmental Impact Assessment (EIA) (85/337/EEC).

In 1992 the European Council adopted the Council resolution on the future Community policy concerning Europe coastal zones in which they concluded that "... there is a clear need for a Community strategy for integrated planning and management of coastal zones based on the principles of sustainability and sound ecological and environment practice" (CEC, 1992). The resolution demanded "a strategy for integrated coastal zone management which will provide a framework for conservation and sustainable use".

Two years later, in 1994, the Council recalled the necessity for a European strategy for integrated coastal zone management and made clear that the Commission should come up in a certain time span with "a view to strengthening coordinated action... [for] a Community strategy for the integrated management of the whole of the Community coastline... " (CEC, 1994).

Consequently, from 1996 to 1999, the EU Commission ran a Demonstration Program on integrated coastal zone management. In total 35 projects contributed to the cross-cutting thematic studies which dealt, for example, with (see EC, 1999): legislation and regulatory instruments, participation, sectoral and territorial cooperation, role of EU policy.

Subsequently, a group of specialists on Coastal Protection (PE-S-CO) was established in 1995. The group was charged to draft a Code of Conduct and to draft a Model Law on Coastal Protection which defines the concept of integrated management and planning based on the principles on sustainable development.

The result was published in January 2000 as "Model Law on Sustainable Management of Coastal Zones and European Code of Conduct for Coastal Zones" (see CEC, 2000). The first part was the proposed Model Law which contained 17 titles dealing with the definition of terms and general principles concerning coastal zones or setting up appropriate bodies to facilitate ICZM. Distinct descriptions were made for each of the 83 articles in the Model Law. The second part contained the Code of Conduct for Coastal Zones that should provide "practical guidelines for the conservation of nature and biodiversity in coastal areas, fully recognizing that socio-economic development in these regions will continue to occur" (CEC, 2000, p. 31). The main intention of the Code of Conduct was to provide practical assistance and guidance to achieve the sustainable development of European coastal areas: "It is intended that this approach will lead to a better dialog within and between the sectors, and with those promoting a more integrated and sustainable form of coastal and marine management and use" (CEC, 2000, p. 32).

In the first part of the Code the principles of the Pan-European Biological and Landscape Diversity Strategy (PEBLDS CEC, 1996) were used to define some key elements for the coastal zone. Each sectoral description was followed by respective guidelines for the sector in coastal areas. Within the Code the process of EIA was seen as a relevant instrument for the sustainable management of coastal zones because it already contained basic and important steps for an integrated management: "The ICZM approach is meant to enhance development and planning models which treat single issues separately, or are implemented by individual administrative units. It is a continuous process active before, during and after sectoral planning" (CEC, 2000, p. 98).

In 2000 the European Commission released a proposal for a European Parliament and Council recommendation concerning the implementation of ICZM in Europe (COM(2000) 545 final) which finally led to the Recommendation on the European Parliament and of the Council of 30 May 2002 concerning the implementation of ICZM in Europe (2002/413/EC). Within six chapters the Commission elucidated the lessons learned from out the projects of the Demonstration Program. The ICZM recommendations were adopted in 2002 (for detailed description, see Sect. 3.2). The important point is that Member States should prepare national ICZM strategies 4 years after the adoption of the ICZM recommendations and that the Commission will propose approximately 1 year later a review of the Recommendations (see EC 2002).

In the following Environmental action program (2002–2012, 1600/2002/EC) emphasis was also laid on ICZM in article 3.10 (... promoting best practice with respect to sustainable land use planning which took into account of specific regional circumstances with *particular emphasis on ICZM program* ...) and in article 6 (2g) (... promoting integrated management of coastal zones ...) in order to achieve sustainable use of the sea and conservation of marine environment.

Next important step for ICZM in the EU was the set-up of the protocol on ICZM in the Mediterranean as additional document to the Barcelona Convention from 1976 (EC, 2009).

Currently, the latest development was the proposal for a Directive establishing a framework for maritime spatial planning and integrated coastal zone management (COM(2013) 133 final, EC, 2013). This action is the result of the adoption of the Integrated Maritime Policy in 2007 (EC 2007a, COM(2007) 575 final) in which the principles of the ICZM recommendations were recalled (see EC, 2007c).

Exercise

- Get the cross-cutting summary reports from the EU ICZM Demonstration Program and read the lessons learned carefully. What are the main statements and findings?
- How are these statements and findings represented within the Model Law and the Code of Conduct?
- Are the Model Law and Code of Conduct helpful producing or supporting national ICZM strategies?
- What could be the barriers why a Law on ICZM has not been further developed?

2.4 ICZM in Selected Member States of the EU

In this chapter selected ICZM strategies of Member States will be introduced. Detailed information on the available ICZM strategies in EU Member States can be found on the website of the European Commission Directorates-General (DG) Environment.[6] An extensive evaluation of European ICZM efforts was conducted in 2006 by Rupprecht Consult and the International Ocean Institute – see Rupprecht Consult (2006) and in Sect. 3.3.2 on p. 43. As stated in the EU recommendations on ICZM reporting periods were set for Member States. The first period was until 2006 and the second from 2006 to 2010. Here, these periods will serve as the basis for descriptions. Three Member States were chosen to indicate the differences in the approach of implementing ICZM. Furthermore, the description of Spain should elucidate the difference between the status planning systems and existing regulations in different EU countries.

2.4.1 Germany

Reporting Period 2002–2006

The national strategy on ICZM for Germany was conducted responsibly by the Federal Ministry of the Environment, Nature Conservation and Nuclear Safety. Germany published the national strategy in March 2006.[7]

The report on the national strategy on ICZM was a stock-take of existing land and sea uses and regulations for the German coast (for example Fig. 2.1), that is, North and Baltic Sea. The first part of the strategy report finished with the description of the status of the coastal environment and the effects of climate change. The second part of the document

Fig. 2.1 View over a salt marsh to the harbor of Bremerhaven, Germany. ©Frank Ahlhorn

[6]For the European Union: http://ec.europa.eu/environment/iczm/home.htm

[7]For Germany: http://ikzm-strategie.de/

was focusing on the description of what ICZM was and what not. The report summarized the ongoing activities on national as well as Federal State level. Finally, the report ended with the question "who is in charge to implement ICZM in Germany?". It was stated that the implementation process should be based on a "top-down" and a "bottom-up" approach. The federal system of Germany led to several laws and regulations on national and Federal State level which are intertwined. Mainly, the national policies are setting the framework for Federal State laws and regulations which include detailed implementation rules. For specific themes only national regulations apply, for example, marine spatial planning outside the 12-nm zone.[8] For other sectors, for example, coastal defense, the Northern Federal States established their own laws and regulations.

Within the national ICZM strategy it was suggested that an ICZM secretariat should further moderate and facilitate the implementation process. Since the publication in 2006 further activities on national and Federal State were conducted, a joint information platform should have been installed, but not a responsible body so far. The ICZM process was accompanied by an ICZM advisory board consisting of different national and Federal State representatives of ministries and authorities.

Reporting Period 2006–2010

Germany prepared a progress report on ICZM according to the ICZM recommendations in 2010. Within the first part of the progress report it was discussed how different activities in several fields contribute to the further implementation of ICZM in Germany. It was mainly concentrated on the EU Marine Strategy Framework Directive and the implementation of maritime EU policies into German law. An important step within ICZM in Germany was the designation of a considerable area in the Economic Exclusive Zone (EEZ) under the NATURA 2000 regulation. The proposed ICZM secretariat has not been fully installed, but the so-called Küsten-Kontor has facilitated the progress in the respective reporting period.

The second part of the progress report contained the description of project-based implementations of ICZM in the German Federal States. Some of these Federal States incorporated ICZM in their spatial planning regulations and laws. An extensive elaboration on several ICZM-related research projects finished this part of the report.

Another part of the progress report contained a compilation of barriers to further implement ICZM in Germany. On the one hand, it stated that European regulations and their intentions have further to be harmonized. On the other hand, obstacles also in the German federal system and the questions about the provision of an additional benefit by ICZM to the existing system of planning, management and laws were drawn. Nevertheless, at the end it was concluded that ICZM is a worthwhile process, but difficult to be implemented.

The progress report ended up with the description of future potential of sustainable coastal development.

---

[8] 1 nm (nautical mile) = 1.852 km.

2.4.2 The Netherlands

Reporting Period 2002–2006

As for the German approach to fulfill the requirement of the reporting according to ICZM recommendations by the EU, the Dutch report[9] was also divided into three parts.

First part of the summarizing document was about the status of the coastal zone. The relevant pressures and uses of the Dutch coastal zone are described. It was stated that the development and preconditions for the use of the Dutch coastal zone differ between the Wadden Sea and the Delta area in the province Zeeland. The contribution, for example, of settlements and facilities for tourism and recreation is more densely located in the Delta area and sandy coast of South Holland. The assessment of the status, on the one hand, and the development of the Dutch coastal zone, on the other hand, was conducted in an integrated way applying the proposed sustainability indicators by the EU ICZM expert group. The main focal point for the status description was the safety against flooding by the sea and subordinate other types of land use. Wherever possible it was striven for the beneficial combination of both safety and ecological, social and economic development, for example, by applying the approach of sand nourishment along the approximately 350 km long sandy coast between Rotterdam and Den Helder (Fig. 2.2).

Within the second part the organization of coastal zone management and policy was described. The administration and management of the Dutch coastal zone was also distributed over several national organizations and institutions as well as in responsibility of the coastal provinces. It was stated that some of the policies and the management were organized according to related sectors, but other sectors were also organized according to the area concerned. For example, until the early 1990s coastal defense was solely striving for the protection of people and infrastructure against flooding based on the recommendations of the Delta Commission I – see Delta Commission (1961) and Correlje et al. (2010). The approach originated from the experiences of the disastrous flooding event in winter 1953. The implementation of the Delta Plan took a long time and is an ongoing process, but the

Fig. 2.2 Beach near the Schelde in the Province of Zeeland, The Netherlands. ©Frank Ahlhorn

[9] For The Netherlands: www.rijkswaterstaat.nl

view on how to safeguard and develop a safe coast, especially in the upcoming light of an integrated management, got broader – see MVenW (1990); TAW (1998); RIKZ (1999); TAW (2000) and Vergouwe (2015).

Since 1990 a set of policies and regulations were developed for the coastal zone in The Netherlands starting with the first policy on coastal areas (Eerste Kustnota) which aimed to achieve a "sustainable flood protection and preservation of functions in the dune areas" (RIKZ, 2005, p. 13) and ending in 2005 with the policy on coastal areas included in the National Spatial Strategy. An important statement which already reflected on the ICZM principles is that reviewing many coastal projects it was striven to solve (urgent) problems in an integrated way taking into account some of the ICZM principles. But none of "the sample projects addressed all of the sectors specified in the ICZM recommendations" (RIKZ, 2005, p. 14).

The third part of the document compiled the important national policies and the general principles for an integrated management of the Dutch coastal zone. The relevant "building blocks" were (RIKZ, 2005, p. 15):

- "the National Spatial Strategy which establishes a national strategy for integrated spatial planning policies in general
- the Third Policy Document on Coastal Areas which provides an integrated framework for coastal zone management and policies on coastal areas."

The implementation of the spatial policy should be achieved by the basic principle of *decentralize wherever possible, centralize only where necessary*. That means that the national government was providing frameworks and guidelines and, thus, the implementation was the task of regional and local governments.

Finally, the document provided some examples of how an integrated approach could be achieved. For example, by sediment-based measures for the sandy coast or by comprehensive stakeholder involvement on different levels in decision-making processes. A suggestion to establish a special institution or organization for integrated coastal zone management in The Netherlands was not specified.

Until 2007 the National Institute for Coastal and Marine Management was responsible for the compilation and preparation of information on (i) sustainable use of estuaries, coasts and lakes and (ii) knowledge for flood risk management. Subsequently, in 2007 the National Institute was dissolved and parts joined the Rijkswaterstaat and other parts were merged with a new institute called Deltares.

Reporting Period 2006–2010

The Netherlands published a progress report on ICZM in 2010 where first of all the new arrangements according to the national governmental bodies were clarified. Ministries were merged and the responsible ministry is now the Ministry of Infrastructure and the Environment where Rijkswaterstaat with the Center of Water Management took over the tasks from the former National Institute for Coastal and Marine Management.

Relevant steps according to the implementation of an integrated strategy in The Netherlands were the introduction of the Spatial Planning Policy Document in 2006, the Coastal Policy Guideline in 2007 and the National Water Plan in 2009.

The National Water Plan was developed for the period of 2009–2015 and substituted the National Spatial Planning Policy Document. The Water Plan incorporated the final recommendations of the Delta Commission released in 2008 (Delta Commission, 2008).[10] Furthermore, the Water Act (went into force in 2006) drew on the allocation of functions and coordination between spatial planning and management of surface and ground water as well as on water defense.

Within this reporting period several communication activities were started not solely concentrated on the coastal zone but on general water issues including the coast, for example, the initiative "The Netherlands living with water."[11]

The final part of the progress report concentrated on the description on ICZM in practice which means that conducted and currently running projects were introduced. These projects mainly focus on the sector of safety against flooding where further sectors or requirements were attached such as nature conservation. For example, by implementing the so-called sand engine in the fore shore zone which was linked to the approach of *building with nature* (Fig. 2.3).

Fig. 2.3 Dunes and holiday houses on the beach at the Schelde, The Netherlands. ©Frank Ahlhorn

[10] www.deltacommissie.com

[11] See e.g. www.helpdeskwater.nl

2.4.3 Spain

Reporting Period 2002–2006

The Spanish report was divided into four parts.[12] The introduction of the Spanish coast was followed by an exhaustive description of the situation today and the national stock-take. The report ends with the elaboration on how Spain wanted to develop and implement an integrated management strategy for the coast (see MERM, 2006).

Within the section "the situation today" the main coastal problems were described in the fields of environment, socio-economic and legal-administrative system. Regarding the environmental problems at the Spanish coast it was stated that urban development and the increase of touristic infrastructure within a one kilometer strip along the coast led to severe degradation of natural coastal habitats and dynamics. Furthermore, some parts of the Spanish coast suffer erosion and low-lying areas adjacent to river deltas also were at risk of flooding. Taking the ecological and chemical status of coastal rivers into account water quality has also to be improved in many water bodies.

The problems mentioned in the socio-economic sector at the coast were linked to urban and touristic development, the fishing and aquaculture and to maritime transport. The touristic sector, for example, concentrated in the recent years mainly on quantity rather than quality. Consequently, the coastal zone suffered increased and, in some places, uncontrolled construction of infrastructure (MERM, 2006, p. 19-20). According to fishing and aquaculture the Spanish fishing fleet had to deal with the entry into force of the UN Convention on the Law of the Sea (UNCLOS). Many coastal states claimed an EEZ under this International law which impedes the free entrance and use of the fish stocks in these areas by foreign fishing fleets.

Within the reporting time Spain had no big maritime logistic hub, although it was recognized that the logistic sector will increase in future time. For the 47 smaller ports along the Spanish coast, administered by national port authorities, the development of a national port strategy was required.

Based on the history of the Roman Law in Spain the coast falls under the status of Public Domain which was defined in the Spanish Constitution from 1978 as follows (Article 132): "The law shall regulate the legal regime of property in the public domain and communal property, based on the principles that such property may not be subject to embargo, divestment and prescription. The areas deemed to belong to the public domain shall be determined by law and shall, in any event, include the maritime-terrestrial zone, beaches, territorial waters and the natural resources of the economic zone and the continental shelf." Ten years later the Spanish Coastal Law defined the above-mentioned areas: "The shores of seas and rias,[13] including maritime-terrestrial zone from low tide to high tide line as well as beaches, dunes, cliffs, marshes and other low-lying wetlands; the terrestrial seas and inland waters, including their beds and undergrounds; the natural resources of the economic zone and continental shelf" (MERM, 2006, p. 10).

---

[12] For Spain: www.magrama.gob.es/es/costas/temas/default.aspx

[13] Estuaries.

According to the right of Public Domain it was difficult to ensure the protection and conservation of the natural environment at the Spanish coast. One of the reasons was that conflicts arose and incompatibility existed with private interests in the coastal zone.

Like other European Member States also in Spain several administrative bodies were responsible for the planning and management within the coastal zone. A complex vertical and horizontal distribution of responsibilities existed. To avoid further disorder of the administration system a general framework for development was needed, especially according to an integrated management at the coast (MERM, 2006, p. 30).

Despite the above-mentioned barriers a variety of networks, organizations and communication platforms existed, for example, the National Commission for Nature Conservation, the Environment Advisory Council or the National Water Council. Within the report it was suggested that a similar cross-sectoral organization should be established for the coast.

Reporting Period 2006–2010

The ambition of drafting a strategy for sustainable coastal management was hardly achieved due to several reasons. These reasons were comprehensively explained in the progress report. One of the reasons should be highlighted in this paragraph:

> By definition, if we are aiming to achieve integrated management of coastal resources, we must reject purely sectoral strategies, which seek to maximize the gains of the sector concerned, even if that causes other sectors with interests on the coast to suffer losses exceeding those gains. The objective of integrated management is to achieve an optimal pattern of exploitation for all sectors. This is where the first difficulty becomes apparent. It is impossible to include all sectors. A decision has to be taken as to which sectors are the important ones and then discussions have to be opened with them to achieve integration, but deciding which are the important sectors is no easy task.
>
> Furthermore, there are sectors that are clearly important but that would prefer to avoid integration so as not to be restricted in their exploitation of a particular resource (this is the case with urban development and land management), other sectors are not important but would like to be considered so and a decision to include a sector as important will always be debatable and disputed. Even when we reach agreement on which sectors are important, including all of them would be totally unmanageable. (MERM, 2010, p. 3)

Consequently, it was concentrated on three, so-called, essential sectors: (i) the public-owned coastal strip, (ii) urban development and town and country planning and (iii) environmental resource. Based on a three-step approach the development of an integrated management approach should be achieved: (i) gathering information and diagnosis, (ii) drafting sectoral plans and (iii) combining sectoral plans to an integrated strategy. These steps were conducted in specified *integrated management units* which are coastal strips with a more or less uniform characteristic (MERM, 2010, p. 4). Step (i) and (ii) were finished to greater or lesser extent, step (iii) was not initialized until 2010. Important steps on the way to an integrated management strategy were cooperation agreements between the responsible national ministry and regional governments.[14]

[14]More detailed description of the results in, for example, Sano et al. (2009).

Despite the fact that Spain ratified the Protocol on ICZM in the Mediterranean (see Sect. 2.3 on p. 23) which might support the efforts of integration the progress report states that "even though completely integrated management of all sectors with interests on the coast may not be achieved, the integration of four, three or two sectors can always be attained, leaving agreements open to other bodies or authorities representing those sectors not initially included to sign up" (MERM, 2010, p. 8).

Exercise

In the respective section references are provided to get more information on the Member States' ICZM strategies. Take a look at the EU and the Member States' websites.

- The EU Recommendations provide a framework for the reports on national ICZM efforts. Compare the three different approaches of Spain, The Netherlands and Germany. What are the similarities, what are the differences?
- What are the main differences according to the problems and challenges the three Member States are facing at their coast?
- According to the integration process what are the major challenges for the planning and management of the respective coastal zone?

References

J.H. Archer and Robert W. Knecht. The U.S. National Coastal Zone Management Program - Problems and Opportunities in the Next Phase. *Coastal Management*, 15:103–120, 1987. doi: 10.1080/08920758709362021.

Bartels. Eigentliche Beschreibung der vor dieser Grafschaft zur See hinaus belegener Eylanden mit anheffter Erinnerung. *Jahrbuch der Gesellschaft für bildende Kunst und vaterländische Altertümer zu Emden*, 4:35–42, 1881.

[San Francisco Bay Conservation and Development Commission] BCDC. San Francisco Bay Plan (reprinted in 2012). Technical Report, San Francisco Bay Conservation and Development Commission, 1969.

[San Francisco Bay Conservation and Development Commission] BCDC. Management Program for the San Francisco Bay. Technical Report, San Francisco Bay Conservation and Development Commission, 1979.

D. Blackbourn. *The Conquest of Nature. Water, Landscape, and the Making of Modern Germany.* W. W. Norton & Company, 2006.

[Council of Europe Committee of Ministers] CEC. Resolution (73) 29 on the Protection of Coastal Areas. Resolution, CEC, 1973.

[Council of Europe Committee of Ministers] CEC. Resolution on the European Coastal Charter. *Official Journal of the European Communities*, 25:124–126, 1982.

[Council of Europe Committee of Ministers] CEC. Council Resolution of 25 February 1992 on the future Community policy concerning the European coastal zone (92/C 59/01). *Official Journal of the European Communities*, C59/1, 1992.

[Council of Europe Committee of Ministers] CEC. Council Resolution of 6 May 1994 on a Community strategy, for integrated coastal zone management (94/C 135/02). *Official Journal of the European Communities*, C135/2, 1994.

[Council of Europe Committee of Ministers] CEC. Pan-European Biological and Landscape Diversity Strategy. *Nature and Environment*, 74, 1996.

[Council of Europe Committee of Ministers] CEC. Model Law on Sustainable Management of Coastal Zones and European Code of Conduct for Coastal Zones. *Nature and Environment*, 101, 2000.

A. Correlje, B. Broekhans, and W. Roos. *Kijk op Waterveiligheid. Perceptie en communicatie van risico's von overstromingen*, chapter Essay 1. Van Deltacommissie 1 tot Deltacommissie 2. Alleen preventie of ook kijken naar overstromingskansen?, pages 10–38. Uitgeverij Eburon, 2010.

[Conference des regions peripheriques maritimes] CRPM. Charte Europenne du Littoral. Charter, CRPM, 1981.

Delta Commission. Eindverslag en Interimadviezen. Deel 1. Technical report, Deltaciommissie, 1961.

Delta Commission. Working together with water. A living land builds for its future. Findings of the Delta Commission 2008. Technical report, Deltaciommissie, 2008.

[European Commission] EC. Communication from the Commission to the Council and the European Parliament on the Integrated Management of Coastal Zones. Technical report, EC, 1995.

[European Commission] EC. Lessons from the European Commission's Demonstration Programme on Integrated Coastal Zone Management (ICZM). Technical Report, EC, 1999.

[European Commission] EC. Directive 2000/60/EC of the European Parliament and of the Council of 23 October 2000 establishing a Framework for Community action in the field of Water Policy. Water Framework Directive, European Parliament and Council, 2000.

[European Commission] EC. Recommendations of the European Parliament and Council of 30 May 2002 concerning the Implementation of an Integrated Coastal Zone Management in Europe (2002/413/EC). *Official Journal of the European Communities*, L184/24:24–27, 2002.

[European Commission] EC. Communication from the Commission to the European Parliament, the Council, the European Economic and Social Committee and the Committee of the Regions. An Integrated Maritime Policy for the European Union. COM(2007) 575 final, European Commission, 2007a.

[European Commission] EC. Directive 2007/60/EC of the European Parliament and of the Council of 23 October 2007 on the Assessment and Management of Flood Risks. *Official Journal of the European Communities*, C288/27:27–34, 2007b.

[European Commission] EC. Commission Staff Working Document Accompanying document to the Communication from the Commission to the European Parliament, the Council, the European Economic and Social Committee and the Committee of the Regions. An Integrated Maritime Policy for the European Union. SEC(2007) 1278, European Commission, 2007c.

[European Commission] EC. Protocol to the Barcelona Convention on Integrated Coastal Zone Management in the Mediterranean. *Official Journal of the European Communities*, L34/19:19–28, 2009.

[European Commission] EC. Proposal for a Directive of the European Parliament and of the Council establishing a framework for maritime spatial planning and integrated coastal management. COM(2013) 133 final, EC, 2013.

[Food and Agriculture Organization of the United Nations] FAO. Integrated coastal management law. Establishing and strengthening national legal frameworks for integrated coastal management. FAO Legislative Study 93, FAO, 2006.

L. Franzius. Die Korrektion der Unterweser (Nachdruck von 1888). *Die Küste*, 51:39–74, 1888/1991.

M.R. Gourlay. History of Coastal Engineering in Australia. In Nicholas C. Kraus, editor, *History and Heritage of Coastal Engineering*, pages 1–88. ASCE, 1996.

N. Harvey and B. Caton. *Coastal Management in Australia*. University of Adelaide Press, 2010. ISBN 978-0-9807230-3-8.

Various authors IPCC. *Preparing to meet the coastal challenges of the 21st century. Conference Report. World Coast Conference 1993*. Ministry for Transport, Public Works, and Water Management, RIKZ/Coastal Zone Management Centre and IPCC, 1994.

J. Kramer and H. Rohde. *Historischer Küstenschutz. Deichbau, Inselschutz und Binnenentwässerung and Nord- und Ostsee*. Verlag Konrad Wittwer, Stuttgart, 1992.

C.C. Mann. *1493. Uncovering The New World Columbus Created*. Alfred A. Knopf, 2011.

[Ministry of Environment and Rural and Maritime Affairs] MERM. Integrated Coastal Zone Management in Spain. Technical Report, MERM, 2006.

[Ministry of Environment and Rural and Maritime Environment] MERM. Report on the Introduction of Integrated Coastal Zone Management in Spain in accordance with the Recommendation 2002/413/EC. Technical Report, MERM, 2010.

W.J. Merrell, M.H. Katsouros, and J. Bienski. The Stratton Commission: The Model for a Sea Change in National Marine Policy. *Oceanography*, 14:11–16, 2001. doi: 10.5670/oceanog.2001.34.

J.F. Mount. *California Rivers and Streams: the Conflict Between Fluvial Process and Land Use*. University of California Press, 1995.

[Ministry of Transport, Public Works and Water Management] MVenW. A new coastal defence policy for The Netherlands. Technical report, Ministry of Transport, Public Works and Water Management, 1990.

[Department of Natural Resources and Environment] NRE. Port Phillip Bay Environmental Management Plan: Background Document. Technical Report, Department of Natural Resources and Environment, 2002.

OECD. Coastal Zone Management. Integrated Policies. Technical Report, OECD, 1993.

[Rijksinstituut voor Kust en Zee] RIKZ. A Coastal Zone Perspective. Preparatory Study. Technical report, RIKZ, 1999.

[Rijksintituut voor Kust en Zee] RIKZ. EU recommendation concerning the implementation of Integrated Coastal Zone Management in Europe. Report on implementation in the Netherlands. Technical Report, Ministry of Transport, Public Works and Water Management, Ministry of Housing, Spatial Planning and the Environment, Ministry of Agriculture, Nature and Food Quality and Ministry of Economic Affairs, 2005.

[Rupprecht Consult and International Ocean Institute] Rupprecht Consult. Evaluation of Integrated Coastal Zone Management (ICZM) in Europe. Technical Report, Rupprecht Consult, 2006.

M. Sano, P. Gonzalez-Riancho, J. Areizaga, and R. Medina. The Strategy for Coastal Sustainability: A Spanish Initiative for ICZM. *Coastal Management*, 38:76–96, 2009. doi: 10.1080/08920750903411734.

J. Sorensen. The International Proliferation of Integrated Coastal Zone Management Efforts. *Ocean and Coastal Management*, 21(1-3):45–80, 1993. doi: 10.1016/0964-5691(93)90020-Y.

J. Sorensen. Baseline 2000 Background Report: The Status of Integrated Coastal Management as an International Practice. Technical Report, Harbor and Coastal Center, Urban Harbors Institute, Univ. Massachusetts, 2000.

[Technical Advisory Committee on Water Defence] TAW. Fundamentals on Water Defence. Technical report, TAW, 1998.

[Technical Advisory Committee on Water Defence] TAW. From Probability of Exceedance to Probability of Flooding. Technical report, TAW, 2000.

[United Nations] UN. Agenda 21 - United Nations Conference on Environment and Development, Rio de Janiero, 3-14 June 1992. Technical report, United Nations, 1992.

G.P. van de Ven. *Man-made lowlands, history of water management and land reclamation in The Netherlands*. ICID and Royal Institute of Engineers in the Netherlands, 1993.

[Victorian Coastal Council] VCC. Victorian Coastal Strategy 2014. Technical Report, Victorian Coastal Council, 2014.

R. Vergouwe. The National Flood Risk Analysis for The Netherlands. Technical report, Ministrie van Vekeer en Waterstaat, Interprovincial Overleg, Dutch Water Authorities, 2015.

[World Commission on Environment and Development] WCED. *Our Common Future. World Commission on Environment and Development/WCED (Brundtlan's Commission)*. Oxford University Press, UK, 1987.

G.C. Wescott. The theory and practice of coastal area planning: linking strategic planning and local communities. *Coastal Management*, 32(1):95–100, 2004. doi: 10.1080/08920750490247535.

3 Status: Basics of ICZM

3.1 Definition of ICZM: Some Suggestions

This section contains a compilation of several definitions of the term *Integrated Coastal Zone Management*. The aim is to indicate the variety of intentions, interpretations and comprehensions of this term. It is not the aim to repeat all available definitions.

The comprehensive description given by the Coastal Zone Management Act of the USA on ICZM has been provided on p. 15 in Sect. 2.1.

Several years after the enactment of the CZM Act in the USA a workshop was held in 1985 attended by an international audience who discussed and finally agreed on the following definition (Sorensen, 1993, p. 49):

> ICZM is a dynamic process in which a coordinated strategy is developed and implemented for the allocation of environmental, socio-cultural, and institutional resources to achieve the conservation and sustainable multiple use of the coastal zone.

Several years later the UN Earth Summit in Rio de Janeiro in 1992 formulated in chapter 17 of the Agenda 21 the term as follows (UN, 1992):

> Coastal States commit themselves to integrated management and sustainable development of coastal areas and the marine environment under their national jurisdiction. To this end, it is necessary to, inter alia:
>
> (a) Provide for an integrated policy and decision-making process, including all involved sectors, to promote compatibility and a balance of uses;
> (b) Identify existing and projected uses of coastal areas and their interactions;
> (c) Concentrate on well-defined issues concerning coastal management;
> (d) Apply preventive and precautionary approaches in project planning and implementation, including prior assessment and systematic observation of the impacts of major projects;

F. Ahlhorn, *Integrated Coastal Zone Management*,
Wasser: Ökologie und Bewirtschaftung, DOI 10.1007/978-3-658-17052-3_3

(e) Promote the development and application of methods, such as national resource and environmental accounting, that reflect changes in value resulting from uses of coastal and marine areas, including pollution, marine erosion, loss of resources and habitat destruction;
(f) Provide access, as far as possible, for concerned individuals, groups and organizations to relevant information and opportunities and participation in planning and decision-making at appropriate levels.

Starting in 1990 the Organisation for Economic Co-operation and Development (OECD) initiated a 3-year study on the management of coastal zones in their member countries; the results were summarized in the report *Coastal Zone Management. Integrated policies* wherein the term of ICZM is defined as follows (OECD, 1993, p. 25):

> Integrated coastal zone management is most simply understood as management of the coastal zone as a whole in relation to local, regional, national and international goals. It implies a particular focus on the interactions between the various activities and resource demands that occur within the coastal zone and between coastal zone activities and activities in other regions. In practical terms this might mean the integration of environmental protection goals into economic and technical decision-making processes, the management of the impacts agricultural run-offs is having on coastal zone water pollution control policies within different parts of the particular coastal zone, or (most probably in practice) all of these and more simultaneously.

Shortly after the UN Earth Summit in 1992, the World Coast Conference was held in The Netherlands in 1993. Here, the term of ICZM was defined as (IPCC, 1994, p. 25):

> Integrated coastal zone management involves the comprehensive assessment, setting of objectives, planning and management of coastal systems and resources, taking into account traditional, cultural and historical perspectives and conflicting interests and uses; it is a continuous and evolutionary process for achieving sustainable development.

In 1996 the Joint Group of Experts on the Scientific Aspects of Marine Environmental Protection (GESAMP) consisting of representatives of different international organizations such as UN, UNEP and FAO and others published a report on *The contributions of science to integrated coastal management* based on several case studies (GESAMP, 1996). Therein, the term Integrated Coastal Management is defined as:

> Integrated Coastal Management (ICM) is a process that unites government and the community, science and management, sectoral and public interests in preparing and implementing and integrated plan for the protection and development of coastal ecosystems and resources. The overall goal of ICM is to improve the quality of life of human communities who depend on coastal resources while maintaining the biological diversity and productivity of coastal ecosystems.

Also in 1996 the World Bank published guidelines as a "conceptual presentation of how Integrated Coastal Zone Management may be applied to contribute to the evolving practice of environmentally sustainable development" (Post and Lundin, 1996, p. v). The definition of the term ICZM is as follows:

> ICZM is a process of governance and consists of the legal and institutional framework necessary to ensure that development and management plans for coastal zones are integrated with environmental (including social) goals and are made with the participation of those affected. The purpose of ICZM is to maximize the benefits provided by coastal zones and to minimize the conflicts and harmful effects of activities upon each other, on resources and on the environment.

The Model Law on Sustainable Management of Coastal Zone defines under Article 2 the term of Integrated Management of Coastal Zones as follows (CEC, 2000):

> "Integrated Management" shall mean sustainable development and use of coastal zones which takes into consideration economic and social development linked to the presence of the sea while protecting landscapes and the coastal zone's fragile biological and ecological balances for present and future generations.
>
> Establishing a system of integrated management of coastal zones necessitates institutional and legislative instruments to ensure participation of the relevant parties and co-ordination of aims, policies and action from both the territorial and decision-making points of view. Integrated management of the coastal zone necessitates dealing with problems not as they arise but in a comprehensive manner, taking account of the interaction between all the elements which make up the environment.

Finally, the definition of ICZM by the European Union laid down in the ICZM Recommendation 2002/413/EC (EC, 2002):

> Member States take into account the sustainable development strategy and the Decision of the European Parliament and of the Council laying down the sixth Community environment action programme, and take a strategic approach to the management of their coastal zones, based on:
>
> (a) protection of the coastal environment, based on an ecosystem approach preserving its integrity and functioning, and sustainable management of the natural resources of both the marine and terrestrial components of the coastal zone;
> (b) recognition of the threat to coastal zones posed by climate change and of the dangers entailed by the rise in sea level and the increasing frequency and violence of storms;
> (c) appropriate and ecologically responsible coastal protection measures, including protection of coastal settlements and their cultural heritage;
> (d) sustainable economic opportunities and employment options;
> (e) a functioning social and cultural system in local communities;
> (f) adequate accessible land for the public, both for recreational purposes and aesthetic reasons;
> (g) in the case of remote coastal communities, maintenance or promotion of their cohesion;
> (h) improved coordination of the actions taken by all the authorities concerned both at sea and on land, in managing the sea-land interaction.

The EU definition provided in the recommendations on ICZM will be the basis for the following chapters of this book.

3.2 Principles for ICZM: A Compilation

One can find many suggestions for the basic principles which should be recognized in designing an integrated management process (see e.g. Fig. 3.1). A bulk of literature is available to substantiate these principles. Here, we concentrate on two descriptions of principles with a time difference of 30 years.

In Sect. 2.1 the first steps in integrated management of the San Francisco Bay were described. In 1972 the Coastal Zone Management Act (CZMA) went into force. Within the CZMA minimum standards have been set for the development of coastal zone management programs of coastal states (Archer and Knecht, 1987, p. 105):

(a) The protection of natural resources, including but not limited to, wetlands, floodplains, estuaries, beaches, dunes, barrier islands, coral reefs, and fish and wildlife and their habitats within the coastal zone;
(b) The management of coastal development to minimize the loss of life and property in hazardous areas;
(c) Priority consideration of coastal-dependent uses, and an orderly process for siting major facilities related to national defense, energy, fisheries development, recreation, ports and transportation, and the location of new development in or adjacent to areas already developed;
(d) Public access to the coasts for recreation purpose;
(e) Assistance in the redevelopment of urban waterfronts and ports, and preservation and restoration of historic, cultural and aesthetic coastal features;
(f) Coordination and simplification of governmental decision-making for the management of coastal resources;
(g) Consultation and coordination with federal agencies;
(h) Participation by the public and local governments in coastal management decision-making; and
(i) Comprehensive planning, conservation, and management for living marine resources, including planning for the siting of pollution control and aquaculture facilities in the coastal zone, and improved coordination between state and federal agencies.

Fig. 3.1 Necessity of integration at the Wadden Sea coast in Schleswig-Holstein, Germany, shipping, fishing and natural conservation © Frank Ahlhorn

In Sect. 2.3 the European developments in ICZM have been delineated. One important policy for ICZM in the EU was the Recommendations on ICZM from 2002 (2002/413/EC). The eight principles are constituted as crucial elements for the development of an ICZM process. The principles are as follows (EC, 2002, p. 25):

(a) A *broad overall perspective (thematic and geographic)* which will take into account the interdependence and disparity of natural systems and human activities with an impact on coastal areas;
(b) a *long-term perspective* which will take into account the precautionary principle and the needs of present and future generations;
(c) *adaptive management during a gradual process* which will facilitate adjustment as problems and knowledge develop. This implies the need for a sound scientific basis concerning the evolution of the coastal zone;
(d) *local specificity* and the great diversity of European coastal zones, which will make it possible to respond to their practical needs with specific solutions and flexible measures;
(e) *working with natural processes and respecting the carrying capacity of ecosystems*, which will make human activities more environmentally friendly, socially responsible and economically sound in the long run;
(f) *involving all the parties concerned* (economic and social partners, the organizations representing coastal zone residents, non-governmental organizations and the business sector) in the management process, for example by means of agreements and based on shared responsibility;
(g) *support and involvement of relevant administrative bodies at national, regional and local level* between which appropriate links should be established or maintained with the aim of improved coordination of the various existing policies. Partnerships with and between regional and local authorities should apply when appropriate;
(h) *use of a combination of instruments* designed to facilitate coherence between sectoral policy objectives and coherence between planning and management.

These principles should build the fundamental basis for all ICZM endeavors in the EU. The challenge regarding these principles is to fulfill them in real and practical life. Many projects were launched to duly complete and to enhance the vitality of these principles.

The enumeration of these principles should be kept in mind for the subsequent chapters, especially for Chap. 3 where the challenges of the integration of sectoral plans and policies are discussed using the example of the management of water-related issues. (see e.g. Fig. 3.2)

3.3 Reflection on the Principles for ICZM

Despite the difficulties of transferring previously mentioned definitions into a processable working environment, also problems and challenges occur in bringing the provided principles into operation. This section provides some reflection on the problems and challenges

Fig. 3.2 After the closure of the Oosterschelde, The Netherlands, in 1986 different demands have to be considered in management, that is protection against high water, nature conservation and leisure shipping. © Frank Ahlhorn

associated with the operationalization of the above-mentioned principles. The focus is laid on the coastal management programs in the USA and the developments of ICZM in the EU.

3.3.1 Reflection on the Coastal Zone Management Act Guidelines

Five years after the CZM Act had passed the Congress the first assessment of the CZM efforts in the USA was conducted. Caused by a missing assessment scheme NOAA (1979)[1] discussed whether "the Act was intended primarily to foster the development of a management process which itself represents the main achievement to be measured or whether the substantive outcomes resulting from these procedures are the measure of accomplishments" (NOAA, 1979, p. 9). On the one hand, in some coastal states of the USA legislation and regulation for the coastal zone already existed. On the other hand, there was a great variety in the coasts of the States regarding planning procedures, the natural environment or the pressures of concern. The attempt of developing an assessment scheme according to the national CZM program highlighted the problem of non-uniform state CZM programs. The CZMA set the frame for the development of State CZM programs; however, it was neither intended to generate a uniform process nor a template for these. In NOAA (1979) the assessment on procedural and substantive results of the state CZM programs was described.

[1] At that time Robert W. Knecht was the Assistant Administrator for Coastal Zone Management at NOAA.

To assess the procedural results NOAA (1979) listed the establishment of legal authorities and other organizations which are linked to CZM within the evaluation period. The substantial results were assessed based on specific indicators (see NOAA, 1979, pp. 19-51). The assessment was enhanced by interviews of several people that show a stake in CZM. These interviews revealed that four aspects were important for the past successes and failures and for the further enhancement of the (national) CZM process:

Comprehensiveness On the one hand, seen as a strength because it pays a tribute to the multifaceted issues of coastal zones. On the other hand, apprehensions were mentioned that taking a lot of aspects into account may lead to "broad-brush" programs owing not enough intention to specific issues. Another important point mentioned was that the provided framework of the CZMA enables profound decision-making. But the individual state programs may not contribute adequately to the overall national intention of, for example, resource management in the entire US coastal zone.

Balancing Some interviewees felt that the CZMA has led to a more balanced decision-making process between, for example, environmental protection and economic development needs. Another strength mentioned was that short-term solution finding processes which may compromise long-term developments were attenuated. But some representatives understood the CZM intentions as fostering and conservation of sector-specific interests.

Costs-savings It was too early to assess the influence on costs neither on project implementation processes nor on administration permits. At this stage both ways cost developments – either increased costs or the complication of processes or raise the effectiveness of coastal decision-making processes – were likely to occur.

Public Awareness The awareness of the broad public increased by the involvement of coastal inhabitants in planning and decision-making processes.

Finally, for the 5-year assessment was concluded that "encouraging signs are appearing that coastal development is being managed more rationally than before, though much more remains to be done in this area" (NOAA, 1979, p. 60).

In a certain period of time the efforts of development and enhancement of CZM efforts drift apparent because of lack of support by the national CZM administration. The process considerably damped down in the 1980s so that a new impulse seemed to be necessary (e.g., see Archer and Knecht, 1987). In these years the role and the activities of the federal program were perceived as inadequate. Recommendations had been suggested to integrate an emerging broader scope for coastal management, that is, the integration of ocean and marine management caused by the extension of the administrative borders at sea (see UN Law of the Seas), and a revised role of the national CZM program administration (see Archer and Knecht, 1987, pp. 109ff).

A comprehensive program evaluation was conducted between 1995 and 1997 by Hershman et al. (1999). The *National Coastal Zone Management Effectiveness Study (CZME)* focused on the five core objectives of the CZMA:

- protection of natural resources such as wetlands and estuaries
- protection of beaches, dune, bluffs,[2] and rocky shore resources
- public access to the coastal zone
- revitalization of urban waterfronts
- seaport developments

First, they identified over 60 studies and reports dealing with the assessment of CZM program on different topics and levels. Hershman et al. (1999) conducted an effectiveness evaluation, so that in their view "effective coastal programs [...] show a clear link between the goal they deem important, the process they set up to achieve that goal and the outcomes resulting from those processes that advance the goal" (Hershman et al., 1999, p. 118). To be able to assess the different core objectives according to the effectiveness they developed and applied three indicator categories: (i) issue importance indicators, (ii) process indicators and (iii) outcome indicators. Each set of indicators was adapted to the respective investigation of the core objectives. Because of the difference of these objectives and the quality and amount of available data approaches have to be tailor-made. In a second step the results and findings on each core objective were aggregated to gain results from the national perspective. The investigation was enhanced by interviews of senior program managers to get an in-depth insight into the development of CZM programs over time. Quintessential statements were, for example, that the senior managers

> "...stressed CZM's unique role in upgrading capacity for management in state and local government units. [...], they emphasized the crucial importance of public participation and the availability of discretionary funding for implementation and specific projects,"
> "...stressed the inadequacy of the decision process prior to CZM and the importance of this institutional restructuring.",
> "...observed that a major outcome of CZM is that only well-conceived development projects are proposed, that is, that CZM standards and requirements help to weed out the poor projects before they reach the stage of seeking official approval"
> "...were forthcoming about failures. A number of them commented that issues like water quality protection, watershed management, and non-point pollution control are not yet well addressed in the coastal zone."

Finally, Hershman et al. (1999, p. 127) concluded that "State CZM programs are effectively implementing the five CZMA objectives examined. However, this conclusion was based on assessment of the policies, processes, and tools used, and on only limited outcome data and case examples that were available." This conclusion was, for example, seconded by the emphasis of the senior managers that "the role of changed governance

[2] Cliff or very steep bank.

arrangements and improved decision processes as a major outcome of the CZMA." The second conclusion touches the quality and availability of data and information and the further development of indicators to be able to adequately assess the effectiveness. The third conclusion was the initialization of national outcome monitoring and performance evaluation system.

3.3.2 Reflection on the Eight ICZM Principles of the EU Recommendations

The assessment of the first period for the development of ICZM strategies and the respective Member States stock-take was conducted by Rupprecht Consult (2006). The comprehensive assessment is based on literature reviews and intensive information collection phase as well as on interviews of responsible organizations and institutions as well as several stakeholders. The distribution of the questionnaire was supported by certain coastal associations in the Member States. Despite these high efforts the feedback was very limited so that a reliable statistical evaluation was impossible. The questionnaire contained also open questions which provided valuable input for a qualitative assessment.

The following paragraphs summarize the main findings of the first assessment process of the implementation of the EU ICZM Recommendation.

In late 2006, 18 of 24 coastal Member States delivered their national reports to be assessed by the evaluation consortium. The overall impression of the status in 2006 was that (Rupprecht Consult, 2006, p. 9):

- no country has implemented an ICZM national strategy as prompted by the EU ICZM Recommendation
- in seven countries [...] the implementation of an ICZM strategy is pending
- in six further countries [...] documents considered as equivalent to an ICZM national strategy have been developed, or coastal zone management strategies become (or planned to become) an integral part of its spatial planning processes
- in eleven countries [...] no ICZM equivalent policies are in advanced stages of preparation, only fragmented tools are in place to address coastal issues.

Based on the interpretation of the qualitative part of the questionnaire and the interviews some pros and cons were featured. Beneficial effects of the EU ICZM Recommendation were, for example, "that new awareness and a high level of preparedness at the regional level regarding long-term coastal challenges" (Rupprecht Consult, 2006, p. 10) have been created. It also led to the revision of the traditional planning process to enable sustainable development. Furthermore, participatory action due to existing regulations already applied received a boost by the execution of the EU Recommendation. It was also acknowledged and, thus, proven that "ICZM [...] could become the instrument to link *terrestrial* to marine legislation, especially on a *Regional Sea level*" (Rupprecht Consult, 2006, p. 10). Aspects which have to be improved were, among others, a better cooperation within Regional Seas, augmented stakeholder involvement and the desirable improvement of monitoring by joint methodologies and long-term funding of local ICZM efforts.

According to the assessment of the ICZM efforts in the North Sea Region all bordering Member States are equipped with a thorough toolkit of spatial planning methods. It was stated that some principles were already in place and some could be improved such as the application of adaptive management approaches. The main findings were summarized as follows (excerpt from Rupprecht Consult, 2006, p. 12–13):

- One of the key obstacles to ICZM is the current strong legislative separation between land and sea-based activities in many of the North Sea countries [. . .]
- Regional Sea Partnerships of the key bodies such as National coastal forums could have a role by facilitating stakeholder participation and dialog in any future system of marine spatial planning.
- Address the problems of consistency, compatibility and accessibility of data collection and storage methods, as well as agreements on cross-border sharing of information in Regional Sea context.
- Develop a set of sustainability indicators that is regularly assessed on the basis of careful monitoring of the coastline and other information could provide the basis for a regular national reporting system to the EU, based as far as possible on data which are simple to collect.
- Use synergies between ICZM and the Water Framework Directive principles (e.g. public participation as key to ICZM and a requirement to WFD and the use of existing coastal observations).

Most Member States at the Atlantic have not already implemented an ICZM strategy. Also a cross-border strategy for the entire Atlantic coastline was missing at that time. Rupprecht Consult (2006) stated that some ICZM principles were recognized in national policies and some not. The assessment for the Atlantic coastal Region was as follows (Rupprecht Consult, 2006, p. 14–15):

- No ICZM strategy has been implemented formally in the five countries. Only first steps have been taken. Spain has targeted the full implementation for 2008.
- Most strategy papers show clearly that the horizontal and vertical flow of information and participation has been neglected in former policies.
- There is a gap between theory and practice in meeting the principles of good ICZM in the countries' strategies. Several countries give the principles as goals of their ICZM, but the reports show that especially participation and communication have not been applied during the development.

Spain belongs to a great extent to the Mediterranean Region so that the main findings for this region will also be briefly touched (Rupprecht Consult, 2006, p. 14–15):

- The most pronounced common problem to the majority of the countries along the Mediterranean is the artificialization of the coast driven by an ever expanding tourism: urban sprawling, building up of second homes, sealing of soils, etc. [. . .]
- There is a multitude of laws, however, a consistent set of laws directing coastal governance and management is usually lacking. The main legislative and policy frameworks governing

> the development in the coast are usually planning instruments that have a physical preponderance and little room for needs of integration of different sectors and participation of stakeholders.

Besides the investigations on success factors and shortcomings of the respective ICZM strategy in each Member State (and the concentrated summary per Regional Sea) in the evaluation report reasons for the differences in the progress of the implementation process were mentioned. Nevertheless, the obstacles and barriers were obvious, but a few aspects have been found that ICZM delivers also an added value in the context of policies and legislation.

Finally, it was discussed whether an incentive approach or a Directive should be taken to foster and improve the implementation of ICZM in the EU, "[...], this evaluation concludes that the potentials of the current EU ICZM Recommendation are not fully exploited, and that an incentive-based approach will be more effective on the EU level" (Rupprecht Consult, 2006, p. 19). The following list indicates the themes where improvement shall be achieved: strengthen the European dimension of ICZM based on a Regional Seas approach, raise the profile of ICZM and enhance its integration with sectoral policies, elaborate the strategic approach of ICZM (oriented at a balanced ecological, social, economic and cultural development) and address major long-term risks: vulnerability to disasters and climate change, endorse awareness, guidance, training and education, enhance stakeholder coordination and participation, perform a mainstreaming of European policies, harmonize monitoring and evaluation frameworks, and improve the knowledge basis for ICZM.

In 2007 the European Commission reported to Parliament and the Council on the evaluation of ICZM in Europe (EC, 2007). This report was based on three different sources; amongst them one was the assessment report of ICZM. "A key achievement of the EU ICZM Recommendation has been to codify a common set of principles that should underlie sound coastal planning and management. While the evaluation confirms the relevance of these ICZM principles, the implementation of the EU ICZM Recommendation also reveals varying interpretations and understanding of ICZM across Europe. To foster a more coherent and effective implementation of ICZM, the principles need to be more operational and better communicated" (EC, 2007, p. 5–6).

Further descriptions are touching the different results and findings such as the need for joint action in monitoring or the necessity for improved cooperation and knowledge exchange between Member States, especially in Regional Sea areas. Furthermore, it is stated that the EU has to support the existing and emerging efforts of ICZM by a variety of instruments: "The Commission will continue to endeavor that these policies and instruments are coherent, so that the implementation at the lower local level of governance is facilitated. Since 2001 [...] the Commission has implemented a range of structured measures to improve openness, participation, accountability, effectiveness and coherence in decision making and implementation of EU policies" (EC, 2007, p. 8).

Finally, the report closes with the following directions to promote ICZM in Europe (excerpt from EC, 2007, p. 9):

- To achieve a more coherent understanding and implementation of ICZM across Member States, guidance needs to be developed to clarify the principles underlying sound coastal planning and management and ways to operationalize them;
- given the high vulnerability of coastal zones to risks and possible impacts related to climate change, strategies to adapt to these risks should be developed and implemented in full coherence with ICZM strategies and instruments dealing with specific natural or technological hazards;
- more efforts are needed for comparative analyses and the communication and promotion of good practices regarding ICZM, including between coastal regions. The gathering of relevant data and effective information sharing and -use in policy and decision-making also needs to be furthered. The development of common indicators and a framework to assess the effectiveness and efficiency of ICZM will need to be continued.

A European Working Group was set-up in 2008 to discuss the performance of the EU ICZM Recommendations (see EC 2009). The report of this Working Group provides a condensed summary of beneficial and improvable aspects of the EU ICZM Recommendation and the process of reporting of Member States. It was their task to suggest a follow-up of the ICZM Recommendation from 2002. Members of the Working Group came from different countries and from different levels of administration and institutions. The discussed policy options as follow-up of the EU ICZM Recommendation were as follows (EC, 2009, p. 7-8):

Revised Recommendation: "The Recommendation is likely to allow a more comprehensive and ambitious setting of the scope and objectives, compared to options based on legally binding instrument. The lacking binding character may be a weakness though to support effective implementation over a longer time-span."

Framework Directive: "The Directive is binding upon each Member State as to the result to be achieved but leaves to national authorities the choice of form and methods."

Decision: "[...] a form of programme at EU level, which could take the form of Decision (e.g. 6th Environmental Action Programme). Emphasis in this option would be on collective actions and commitments to support ICZM, rather than a more detailed framework for implementation of ICZM in and by Member States."

Regional Sea Convention: "Working through [this instrument] may be an option to address in particular the different regional contexts. [...] The type of instruments available also varies among conventions (protocol, recommendation, action plan). This option should therefore best be considered as a complement to the other options above."

Exercise

- Compare the Hershman et al. (1999) recommendations with the Rupprecht Consult (2006) recommendations. What are the similarities? What are the differences? To what extent do you guess that the different constitutions of the USA or the EU are influential to your findings?
- Read the success as well as the failed factors mentioned in Rupprecht Consult (2006) carefully. Compare these factors from 2006 with the current situation in selected EU Member States. For example, what about the performance of selected Mediterranean Member States according to the agreement on the Barcelona Protocol on ICZM in 2009?
- Rupprecht Consult (2006) touched the question whether the further implementation process of ICZM should be incentive-based or regulated by an EU Directive. Discuss the pros and cons of an incentive-based or a mandatory approach.
- One of the statements in the report of the EU Commission (EC, 2007, p. 9) was: "... strategies to adapt to these risks should be developed and implemented in full coherence with ICZM strategies and instruments ...". Most of the ICZM strategies of the Member States have been developed on the basis of existing rules, laws and directives. One should suppose that these regulations already consider different types of risk in one or the other way. If not, how can the relationship between sectoral policies and the ICZM strategy according to the handling of risk make a difference?
- Given the brief description of the conclusions for the follow-up to the EU ICZM Recommendation by the Working Group in 2009 and the listing of ICZM efforts. Find Regional Sea Conventions in Europe and describe in which way ICZM is considered. Draft a comparison between the EU Recommendation and the achievements of the Regional Sea Conventions according to ICZM efforts.

References

J.H. Archer and Robert W. Knecht. The U.S. National Coastal Zone Management Program - Problems and Opportunities in the Next Phase. *Coastal Management*, 15:103–120, 1987. doi: 10.1080/08920758709362021.

[Council of Europe Committee of Ministers] CEC. Model Law on Sustainable Management of Coastal Zones and European Code of Conduct for Coastal Zones. *Nature and Environment*, 101, 2000.

[European Commission] EC. Recommendations of the European Parliament and Council of 30 May 2002 concerning the Implementation of an Integrated Coastal Zone Management in Europe (2002/413/EC). *Official Journal of the European Communities*, L184/24:24–27, 2002.

[European Commission] EC. Report to the European Parliament and the Council: An evaluation of Integrated Coastal Zone Management (ICZM) in Europe. COM(2007) 308 final, European Commission, 2007.

[European Commission] EC. Report from the Working Group follow-up to the EU ICZM Recommendation. Technical Report, EC, 2009.

[Joint Group of Experts on the Scientific Aspects of Marine Environmental Protection] GESAMP. The Contributions of Science to Integrated Coastal Management. Reports and Studies 61, IMO, FAO, UNESCO-IOC, WMO, WHO, IAEA, UNEP, 1996.

M.J. Hershman, J.W. Good, T. Bernd-Cohen, R.F. Goodwin, V. Lee, and P. Pogue. The Effectiveness of Coastal Zone Management in the United States. *Coastal Management*, 27:113–138, 1999. doi: 10.1080/089207599263811.

Various authors IPCC. *Preparing to meet the coastal challenges of the 21st century. Conference Report. World Coast Conference 1993*. Ministry for Transport, Public Works, and Water Management, RIKZ/Coastal Zone Management Centre and IPCC, 1994.

[National Oceanic and Atmospheric Administration] NOAA. The First Five Years of Coastal Zone Management. An Initial Assessment. Technical Report, NOAA, 1979.

OECD. Coastal Zone Management. Integrated Policies. Technical Report, OECD, 1993.

J.C. Post and C.G. Lundin. Guidelines for Integrated Coastal Zone Management. ESD Studies and Monographs, No. 9, The World Bank, 1996.

[Rupprecht Consult and International Ocean Institute] Rupprecht Consult. Evaluation of Integrated Coastal Zone Management (ICZM) in Europe. Technical Report, Rupprecht Consult, 2006.

J. Sorensen. The International Proliferation of Integrated Coastal Zone Management Efforts. *Ocean and Coastal Management*, 21(1-3):45–80, 1993. doi: 10.1016/0964-5691(93)90020-Y.

[United Nations] UN. Agenda 21 - United Nations Conference on Environment and Development, Rio de Janeiro, 3-14 June 1992. Technical report, United Nations, 1992.

Hydrology of (Shallow) Coastal Regions

4

Helge Bormann

4.1 The Water Cycle of Coastal Areas

The water cycle of coastal regions is characterized by the interplay between the different hydrological processes. It can be described by the general water balance equation of a system, quantifying the change of water in a defined area:

$$\Delta S = P - ET + Q_{\text{in}} - Q_{\text{out}} \tag{4.1}$$

where ΔS = change in water storage, P = precipitation, ET = evapotranspiration, Q_{in} = inflow, Q_{out} = outflow.

Looking at a catchment, defined by the absence of lateral inflow across the catchment boundaries, Eq. 4.1 reduces to

$$P = ET + Q + \Delta S \tag{4.2}$$

where Q = runoff. This means that precipitation splits up in evapotranspiration and runoff, and can replenish or reduce the catchment storage.

Due to the humid climate, many coastal regions have a positive water balance, for example, the North Sea Region in central Europe. In case of a positive water balance, the precipitation P exceeds the evapotranspiration ET. The excess water either generates runoff or is temporally stored in the hydrological system (Fig. 4.1).

In the neighborhood of estuaries, which are defined as partially enclosed coastal body of brackish water with one or more rivers flowing into it, and with a free connection to the open sea, and tidal rivers, coastal regions are also influenced by rivers conveying the excess water of inland catchments to the sea. High flows, inducing high water levels, can lead to inundations while low flows accelerate the drainage of the respective area.

F. Ahlhorn, *Integrated Coastal Zone Management*,
Wasser: Ökologie und Bewirtschaftung, DOI 10.1007/978-3-658-17052-3_4

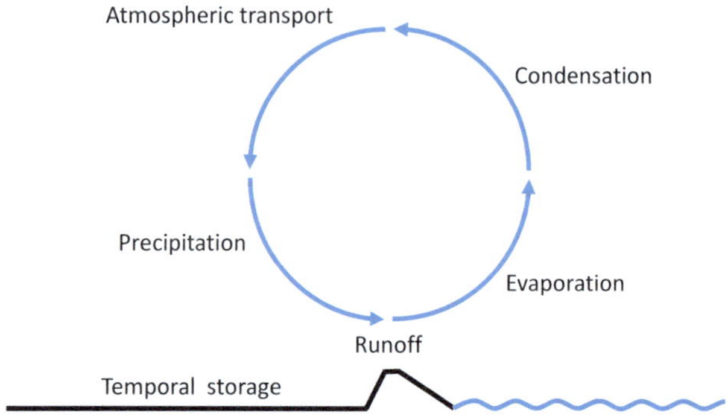

Fig. 4.1 Schematic water cycle of humid coastal areas. Precipitation exceeds evapotranspiration on the land side, resulting in a positive water balance. Excess water flows toward the sea

Shallow coastal regions mostly are also characterized by a flat terrain. A flat terrain favors slow hydrological flow processes compared to quick flow processes, for example, related to runoff generation. Usually, in flat terrain subsurface flow processes are more important than surface runoff, except in case of soil saturation. Then, water accumulates at the surface and may generate surface runoff.

In addition to the topography, dominance of hydrological processes also depends on the soil properties and on the soil composition of coastal regions. While fine-textured soil material tends to a better storage behavior, coarse-textured soils have a higher water permeability. Storage capacity and permeability directly have an impact on the runoff generation processes.

Figure 4.2 shows the spatial distribution of the permeability at the Northwest German North Sea Coast.[1] Large parts of the shallow coastal areas are dominated by soils with a low permeability. The fine-textured marsh soils mainly consist of clay, silt and organic matter. They are characterized by a low permeability and a high productivity. For agricultural (or any other anthropogenic) use they require an intense drainage (see the dense drainage network in Fig. 4.2).

The implications of these delineated hydrological characteristics for the water management of coastal regions are as follows (see also Figs. 1.3 and 1.4):

- Humid coastal regions are characterized by a positive water balance. If these regions shall be used for agriculture or settlements, the area needs to be drained, and the drained water needs to be conveyed to the sea,
- Shallow coastal regions are characterized by a flat terrain. Therefore, hydraulic gradients are small, and undrained coastal regions are usually wet landscapes. Efficient drainage of such regions requires methodologies to accelerate water flow in order to keep the regions dry.

[1] k_f is the coefficient of permeability or hydraulic conductivity; data has been provided by the Lower Saxonian Agency for Mining, Energy and Geology.

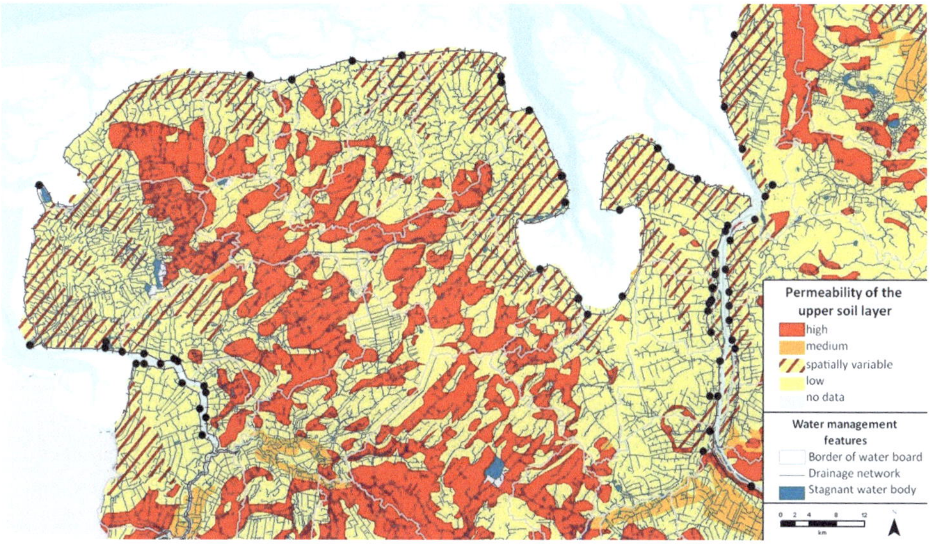

Fig. 4.2 Permeability of the upper soil layer along the Northwest German North Sea coast. High = $k_f < 1 * 10^{-5}$ m/s, medium = $1 * 10^{-5}$ m/s $\leq k_f < 1 * 10^{-4}$ m/s and low = $k_f > 1 * 10^{-4}$ m/s, spatially variable = highly variable soil layer unable to allocate a specific permeability class. Source: adapted from Ahlhorn et al. (2010)

4.2 Impacts of the Sea on Coastal Fresh Water Systems

The tidal oscillation is an important boundary condition influencing the water cycle of coastal areas. Caused by the gravitational forces of the earth, the moon and the sun, the sea water level periodically increases and decreases approximately shaped like a sine wave. The gravitational attraction of moon and earth governs the rhythm of the oscillation due to their small distance compared to the distance between sun and earth. Therefore, the lunar orbit determines the phase and the frequency of the oscillation of the sea water level. Along most of the coastlines, the periodic time is 12:24 hours. The amplitude of the oscillation depends on the inference of the lunar tide with the solar tide (spring tide when both tidal effects are adding up, versus neap tide when the solar tide is subtracted from the lunar tide). The tidal forces have stronger impacts along coastlines compared to the high seas, and the tidal range is strongly influenced by the shape of the coastline. The average amplitude of the oscillation (the tidal range) ranges from several decimeters until more than 10 m (Kelletat, 1989).

Figure 4.3 exemplifies a period of two days of a tidal gauge located in the North Sea (tidal gauge Leyhörn). The tidal range is about 3 m, oscillating around mean sea level with an oscillation period of around 12.5 hours. The consecutive tidal maximum (as well as the minimum) reach approximately the same water level.

Fig. 4.3 Tidal oscillation of the water level in the North Sea at gauge Leyhörn

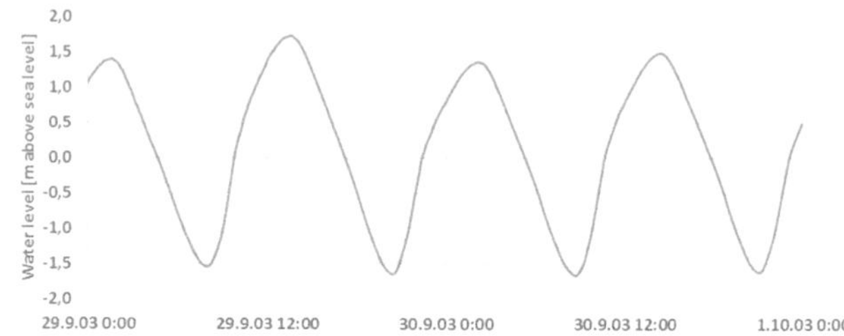

In addition to the diurnal tides, storm tides are a key risk for coastal areas. In periods of strong onshore winds, the high tides can be significantly raised by the wind. If such storms persist several tides, each high tide accumulates more water, resulting in further elevated high water levels. Highest storm tides happen if such storms develop during spring tides. In the North Sea Region, the storm season usually lasts from October to February. Storm tides can be classified through their return period or accordingly the probability of exceedance; see, for example, Baumgartner and Liebscher 1996; Gönnert et al. 2001): moderate (less than 2-year return period), severe (2–20-year return period) and very severe (more than 20-year return period).

Figure 4.4 provides an example for accumulating tidal maximum during a storm period from October 30 until November 1, 2006, in the North Sea. The tidal diagram form from Leyhörn shows a significant increase in the maximum water levels of four subsequent high tides. Due to the storm, pressing the sea water into the German Bight, water cannot retreat during the low tide phase, finally resulting in an extreme water level approximately 38 hours after the first maximum shown here.

Another important factor is the different composition of sea water compared to fresh water. While sea water usually consists of salt water (salt concentration exceeding 1.8 %), most inland water systems consist of fresh water (salt concentration remaining below 0.05 %). The North Sea, for example, has a salt concentration of 3.4–3.5 %.

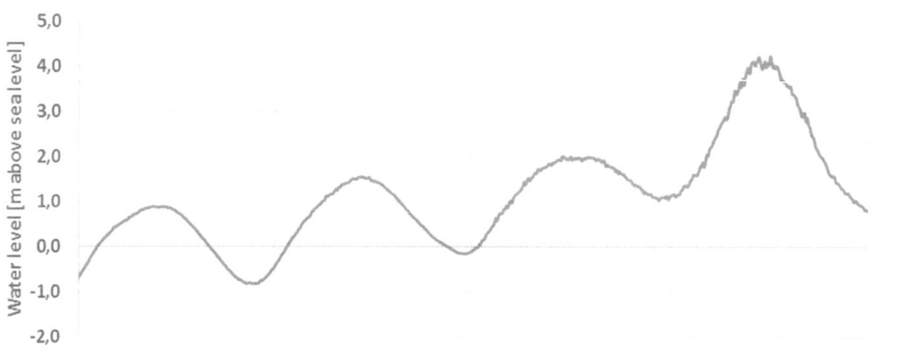

Fig. 4.4 Tidal oscillation of the water level in the North Sea at gauge Leyhörn during the storm tide in November 2006

Salt concentration influences the specific gravity of water. It can therefore cause a stratification of different waters with different salinities.

What impacts do water-level fluctuations and differences in composition have on the hydrological processes of shallow coastal regions?

4.2.1 Impacts on Water Quantity

Tidal Influence on Coastal Rivers

The oscillation of the sea water level leads to a fluctuating gradient of the water level in the estuaries, influencing the flow velocity of coastal rivers. During high tides, river water is impounded, and if the sea level exceeds the water level of the river, sea water can even flow upstream (see Sect. 5.4). Rivers at shallow coasts can therefore also experience a change in flow direction: while during low tide water flows downstream, water can flow upstream during high tides (Fig. 4.5). The described flow pattern is superimposed by a mass density-driven flow caused by the different mass densities of salt water and fresh water. The mass density-driven flow leads to an upstream salt water flow at the river bed while fresh water flows downstream at the water surface. The result is that during slack water conditions, meaning that the net water flux in the river or estuary is zero, upstream and downstream fluxes of salt and fresh water may occur (Fig. 4.5).

Impoundment and upstream water flow lead to an oscillation in estuaries and coastal rivers. That phenomenon can be observed in the hinterland up to the tidal limit, which is the spatial boundary of tidal influence. There, the tidal oscillation of the surface water level is similar to that at coastal gauges (Fig. 4.3) while the morphology of the estuary may induce asymmetries of the rising and falling limbs.

The propagation of the tides in estuaries and coastal rivers can be described by tidal waves, moving upstream into the estuaries. Tides therefore are delayed at upstream sites compared to the coastline. In case of funnel-shaped estuaries and rivers, the amplitude of this wave (tidal range) can even increase upstream, as observed for the Elbe and Weser rivers in Germany, for example. Dredging of rivers and estuaries can amplify such effects (see Sect. 5.4).

Ground Water Connected to Tidal River or to the Sea

Generally, surface water and shallow ground water systems are interacting. Changes in river and/or sea water levels therefore also induce changes in the water levels and the dynamic of shallow ground water systems. Such influence is strongest close to the river and the coastline, and it decreases with increasing distance from the surface water body (Fig. 4.6). The range of such ground water oscillations in the neighborhood of tidal rivers is proportional to the ground water flow velocity. According to Darcy's law,[2] the velocity

[2]Darcy's law in general: $v = k_f * I$, where v = rate of water flow (volume per time), k_f= coefficient of permeability and I= hydraulic gradient.

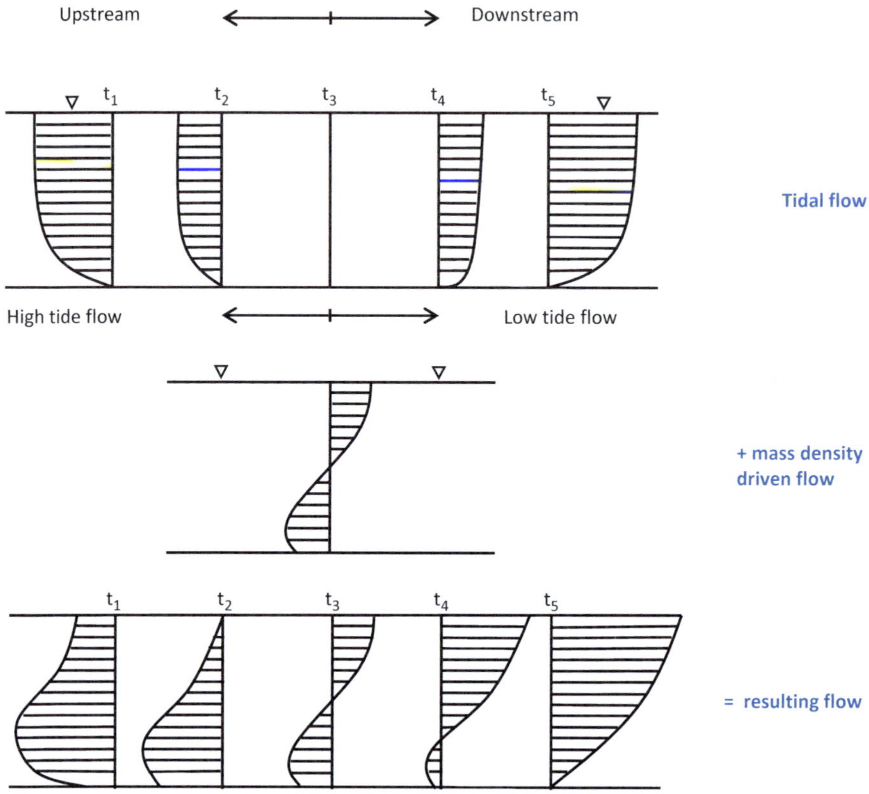

Fig. 4.5 Impact of tides and mass density on the flow direction and flow velocity distribution in estuaries. While tidal flow is only governed by the gradient in the water level (*top*), the mass density driven flow leads to an upstream salt water flow at the river bed while fresh water flows downstream at the water surface (*center*). The superposition of both flows (*bottom*) is the resulting flow pattern in the transition zone (t_1 = high tide; t_2 = intermediate stage; t_3 = slack water; t_4 = intermediate stage; t_5 = low tide). Source: adapted from Baumgartner and Liebscher (1996)

is calculated as product of the saturated hydraulic conductivity and the gradient of the hydraulic head. Hence, the range of tidal influence of ground water systems is largest for coarse sediments and large tidal ranges.

Low-Lying Areas (Marsh Land)

Low-lying flat areas directly connected to the tidal rivers or to the sea are regularly inundated. Comparable to the Wadden Sea (Fig. 1.1), inter tidal flats can also develop in the hinterland (Fig. 4.7). While such processes usually are prevented through artificial drainage and flood protection measures, these processes are the origin and nature of the landscape. Only due to regular inundation over a long time period the marsh landscape could develop in historical times.

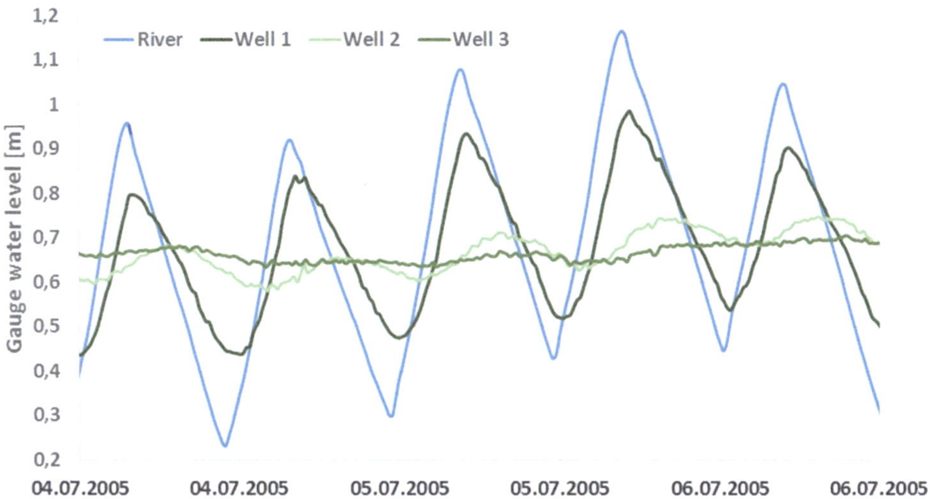

Fig. 4.6 Ground water level oscillations at three wells close to a tidal river (Ollenbäke, Germany): 5 m (well 1), 10 m (well 2) and 15 m (well 3) distance to the river

Fig. 4.7 Inter tidal inland areas recently formed by water-level oscillations of tidal river (Ollenbäke, Germany). © Helge Bormann

4.2.2 Impacts on Water Quality

Estuaries and Coastal Rivers

Estuaries and coastal rivers are the locations where salt and fresh water meet. The transition zone between fresh water in the rivers and salt water in the ocean is typically located in the estuaries. Due to mixing processes, the transition zone consists of brackish water. As mentioned above, fresh water has a smaller specific gravity compared to salt water inducing a stratification of the different waters: fresh water is located at the rivers surface, moving downstream, while salt water is close to the river bed, moving upstream like a salt water wedge (Fig. 4.8). In case of a laminar flow system stratification of the different waters can be observed (fresh water at the water surface, salt water at the river bed, brackish water in between).

The transition zone does not have a permanent location. Driven by the tides and the river flow, the brackish zone moves along the river or accordingly the estuary depending on the sea water level (high versus low water level) and on the river discharge (high versus low discharge). While during high tides, the brackish zone moves upstream compared to low tide, high river discharges lead to a downstream movement of the brackish zone in the estuary compared to low discharges (e.g., see Figs. 5.15–5.17).

Figure 4.9 gives a schematic impression how the tides and the river flow influence the transition zone. The upper part of the figure shows the salt concentration, the lower part the change in salt concentration along the coastal river (or estuary). During low tides (dashed line) the transition zone moves downstream while it moves upstream during high tide (continuous line). The impact of river discharge is vice versa: the transition zone moves downstream if the river discharge is high (light blue), and upstream in case of low river discharge (dark blue). The lateral shift of the curve describing the change in salt concentration can be interpreted as up- or downstream movement of the transition zone in the river or in the estuary.

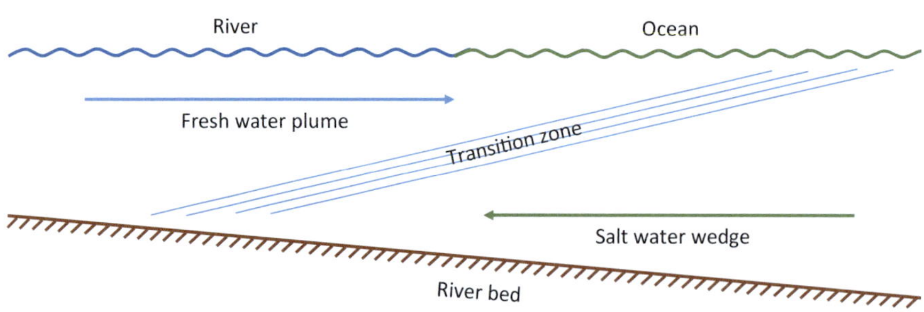

Fig. 4.8 Transition zone of fresh water and salt water in an estuary. Stratification of the different waters due to specific gravity

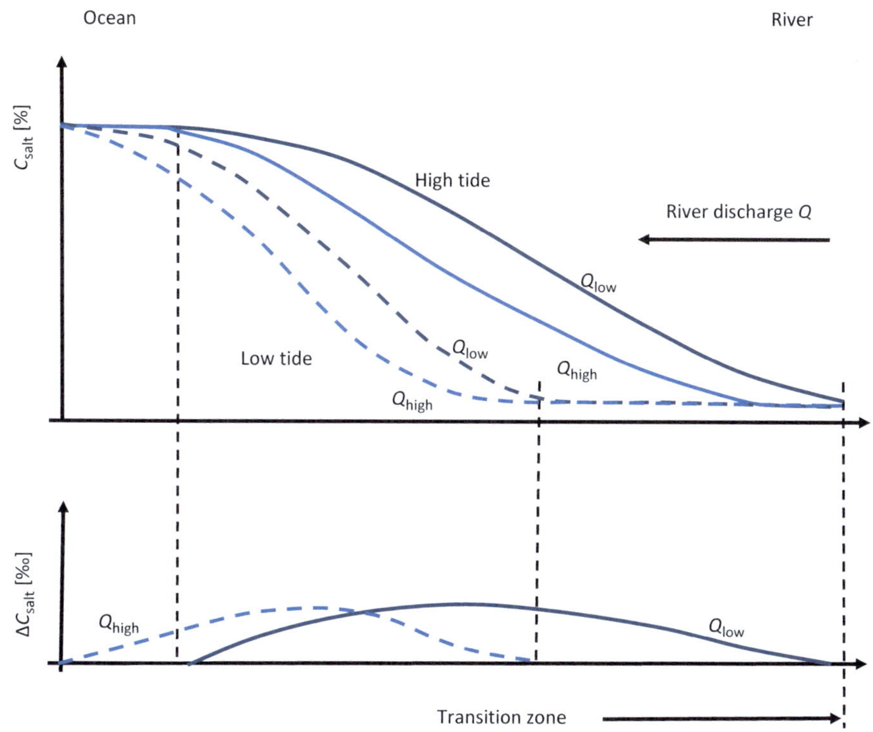

Fig. 4.9 Movement of the brackish water zone in estuaries: During high tides, the brackish zone moves upstream compared to low tide. By contrast, large river flows lead to a downstream movement of the brackish zone in the estuary. C_{salt}: salt concentration; ΔC_{salt}: change in salt concentration; Q_{high}: high river discharge; Q_{low}: low river discharge. Source: changed after Baumgartner and Liebscher (1996)

Ground Water

Coastal and island aquifers are prone to sea water intrusion. Because saline water is denser than fresh water, it invades aquifers which are hydraulically connected to the ocean. Like in the surface water bodies, the transition zone is a zone of mixing between fresh and salt water. Under equilibrium conditions, the interface remains fixed. The transition zone is found in the ground water along the coastline. Since flow processes usually are slow in the ground water, the transition zone is relatively narrow (Fig. 4.10). Intense fresh water drainage or fresh water abstraction on the land side can induce further expansion of the saline ground water landwards.

Islands develop their own fresh water lenses. Under natural conditions, fresh water forms a lens that floats on the top of a base of salt water. Recharge is happening due to infiltration of rainwater into the fresh water lens (Fig. 4.11).

Sea water intrusion must be addressed in managing ground water resources in island and coastal areas. The fresh water lens is very susceptible to contamination,

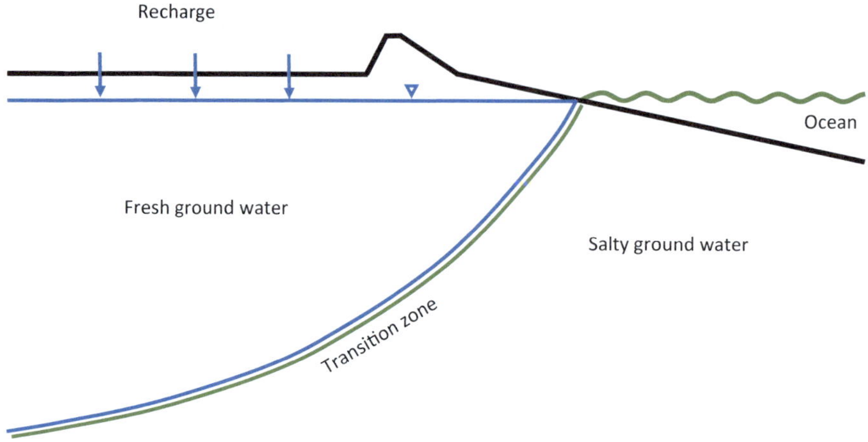

Fig. 4.10 Transition zone of fresh water and salt water in the ground water

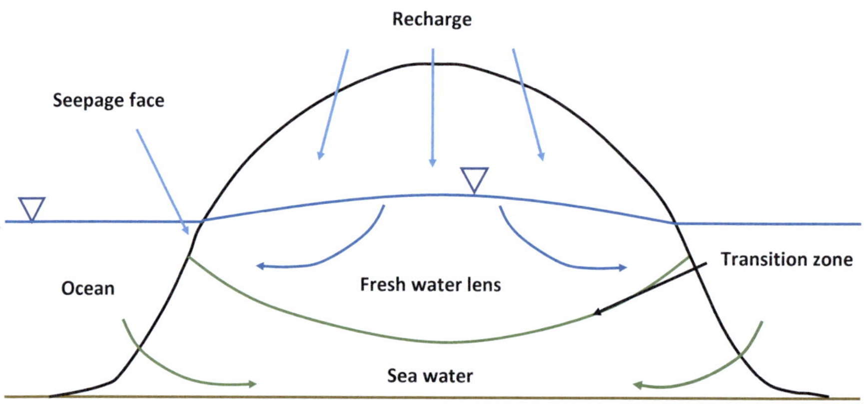

Fig. 4.11 Fresh water lens of an island. Source: changed after Maidment (1993)

and once contaminated, it can be very difficult to restore the pristine conditions (Maidment, 1993).

Specific Impacts of Storm Tides

Storm tides are characterized by extreme water levels, strong winds, high waves and strong forces which affect the coastline. These different factors are the reason for the high damage potential of storm tides. In addition to coastal erosion, inundation of usually dry areas entails a high damage potential.

In this respect fresh water resources are extremely vulnerable to inundations. Inundating salt water pollutes the soil (reducing the agricultural productivity) and can infiltrate into the ground water. As a consequence of salt water infiltrations, ground water resources need decades for regeneration.

4.3 Implications for (and of) Drainage Management

The diurnal oscillations of the water level, including inundations and tidal storm floods, make use of those areas, which are influenced by this dynamic, very difficult. In the past, the human strategy therefore was – and in large parts still is – to prevent the natural hydrological dynamics, including regular inundations and wet conditions, and to control the hydrological processes of such regions. In the North Sea Region, sea walls have been built for 1000 years, and due to the positive water balance, drainage systems have been developed (see Chap. 5).

Implications for drainage management, in direct connection with coastal protection measures, are to protect the respective areas against tides and storm tides while ensuring water drainage through technical solutions. Possible technical solutions are:

- Constructing dikes (sea walls) to prevent the landscape against tidal inundations
- Constructing sluices to convey the excess water from the hinterland to the sea (where topography allows for that, requiring hydraulic gradient from inland water to the sea) (Fig. 4.12 shows the hydrographs of sea and inland water level and highlights the period appropriate for gravitational drainage)

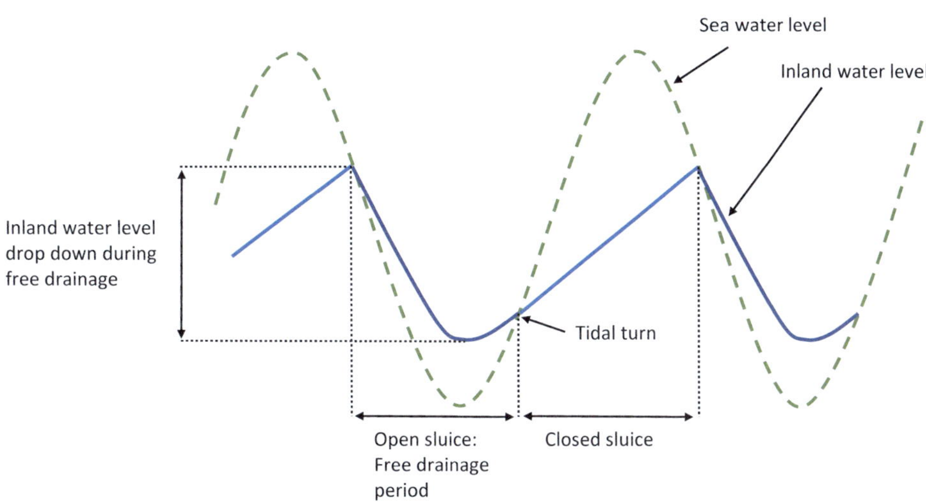

Fig. 4.12 Free drainage periods through sluices depending on the water levels of the sea and the inland waters and on the sluice management (open if inland water level exceeds sea level, closed if sea water level exceeds inland water level). Source: adapted from Riedel (2004)

- Installing pumping stations to convey the excess water from the hinterland to the sea where no natural hydraulic gradient is available
- Constructing storm surge barriers to protect the tidal rivers against high water levels and to reduce the length of the dike system

All of these solutions have been implemented in practice in central Europe for a long time. For further details see Sect. 5.1.2.

Applying a drainage management to an originally wet landscape certainly has implications on the landscape itself and on its water bodies.

The protection against the sea water excludes the natural hydrological dynamics from the related coastal regions. Sedimentation, which was the vital process during the landscape development, does not happen anymore. Salt concentrations in soils and water bodies are reduced by dilution processes caused by rainfall, infiltrating water and ground water recharge. The geochemical environment is changing fundamentally. The aeration of the marsh soils caused by drainage accelerates mineralization processes, which together with de-calcification is the reason for soil subsidence.

The drainage of water bodies leads to adjusted and constant water levels in the water bodies. The water bodies in the marsh area are no longer exposed to the tidal dynamics. While such constant levels are advantageous for anthropogenic use of the landscape, the hydrological nature of the water bodies is suppressed. To ensure the hydraulic function of the water bodies, regular dredging and maintenance is required. The consequence is that, according to the European Water Framework Directive (WFD, EC 2000), such water bodies are classified either as heavily modified water bodies or even as artificial water bodies.

In terms of flood risk the respective areas are protected by dikes, allowing anthropogenic activities and long-term investments. But, however, a 100 % safety through such technical protection systems does not exist. Therefore, according to the European Flood Risk Management Directive (FRMD, EC 2007) these coastal areas are classified as high-risk areas. In case of failure of the protection systems, large areas might be flooded.

In addition to the today's water management challenges, climate change is expected to aggravate an efficient water management (Bormann et al., 2012). Due to a rising sea level, an intensification of the terrestrial water cycle (increase in heavy rainfalls in frequency and intensity) at least temporarily more water will have to be drained against a higher sea level (Fig. 4.13). Similarly, salt intrusion is expected to be more severe than today.

As a consequence, in our days, adaptation to climate change is required, and traditional water management needs to be checked against future drainage requirements and innovative solutions (Bormann et al., 2015).

Exercise

1. Large parts of the coastal areas are protected by dikes. That implies that many inland water bodies are regulated by water management. Which hydrological processes

Fig. 4.13 Future water management needs in shallow coastal regions: Sea-level rise, intensification of the terrestrial water cycle and land subsidence will aggravate an efficient drainage

would be affected if you would open a dike section? Explain, which processe would be affected, and in which way.

2. Most coastal areas are characterised by a humid climate, resulting in a positive terrestrial water balance. What effect does coastal protection have on the water balance of the protected hinterland? In which way does water management compensate for these impacts? Identify the problem and compare different solution options.
3. Climate change is a major issue in coastal regions. Look at the water balance equation: which fluxes and states are directly influenced by climate change, and which ones can be affected indirectly? Explain possible changes based on observed and projected climate change in your region.

References

F. Ahlhorn, J. Meyerdirks, and I. Umlauf. Speichern statt pumpen. Abschlussbericht. Technical report, University of Oldenburg, 2010.

A. Baumgartner and H.-J. Liebscher. *Allgemeine Hydrologie. Quantitative Hydrologie*. Borntraeger, 1996.

H. Bormann, F. Ahlhorn, and T. Klenke. Adaptation of water management to regional climate change in a coastal region – hydrological change vs. community perception and strategies. *Journal of Hydrology*, 454-455:64–75, 2012.

H. Bormann, R. van der Krogt, L. Adriaanse, F. Ahlhorn, R. Akkermans, Y. Andersson-Sköld, C. Gerrard, N. Houtekamer, G. de Lange, A. Norrby, N. van Oostrom, and R. de Sutter. *Handbook of Climate Change Adaptation*, chapter Guiding Regional Climate Adaptation in Coastal Areas, pages 337–357. Springer, 2015.

[European Commission] EC. Directive 2000/60/EC of the European Parliament and of the Council of 23 October 2000 establishing a Framework for Community action in the field of Water Policy. Water Framework Directive, European Parliament and Council, 2000.

[European Commission] EC. Directive 2007/60/EC of the European Parliament and of the Council of 23 October 2007 on the Assessment and Management of Flood Risks. *Official Journal of the European Communities*, C288/27:27–34, 2007.

G. Gönnert, S.K. Dube, T. Murty, and W. Siefert. Global Storm Surges. *Die Küste*, 63:623, 2001.

D. Kelletat. *Physische Geografie der Meere und Küsten*. Teubner Taschenbuch, 1989.

D.R. Maidment. *Handbook of Hydrology*. McGraw-Hill, 1993.

G. Riedel. *Ein hydrologisches Modell für tidebeeinflusste Flussgebiete*. Dissertation, Leichtweiss-Institut für Wasserbau der TU Braunschweig, 2004.

Challenges: Sectoral Descriptions

5

5.1 Coastal Protection Management

5.1.1 Introduction

The story about coastal protection has to be told because more than thousand years ago coastal inhabitants decided and were able to protect themselves against the rising water and floods. Although the definite beginning of building dikes could not be determined, evidence has been found that already in Roman times low mounds of clay have been raised (Bantelmann, 1966; Blok, 1984; Hallewas, 1984; Brandt, 1992). Here, the question of the beginning is not that important, it is more interesting to get an impression of the development on how coastal inhabitants react on the challenges for the protection against flooding caused by changing circumstances.

Allemeyer (2007) compared and discussed different views of persons who lived or worked in the seventeenth and eighteenth centuries at the German coast. The confrontation exemplifies the different mentalities that have already been existing in these times. The concentration was laid, on the one hand, on the relationship between the protection performance of dunes and (artificial) sea walls (i.e., dikes) and, on the other hand, between the ability of human kind of trying to *tame* or *fight* actively against the sea.

Bartels (1881) prepared a description of the situation on the East Frisian islands in 1650. The first part of his report described the situation on several islands, but in the second part he pinpointed on the linkage between the existence of the islands in front of the mainland and the status and efforts for coastal protection at the marsh. He recommended to strengthen and maintain the efforts of the inhabitants of the islands to improve dune stability because it provides positive effects for mainland protection. Bartels called the East Frisian islands *breastwork* for the mainland. The way how to deal with these dunes

F. Ahlhorn, *Integrated Coastal Zone Management*,
Wasser: Ökologie und Bewirtschaftung, DOI 10.1007/978-3-658-17052-3_5

Fig. 5.1 Dunes at the Schelde protecting the hinterland, The Netherlands. ©Frank Ahlhorn

should be learned from the Dutch neighbors and the improvement should be done *with* and *not against* nature (Fig. 5.1).

The same conclusions were drawn by inhabitants at the peninsula Eiderstedt in Schleswig-Holstein. In 1770 the responsible authority wanted to protect the mainland against flooding by building a dike. But the inhabitants insisted to construct an artificial protection wall. Allemeyer (2007) pinpointed that a distinction was made between *natural* and *artificial* coastal protection. Of course, one of the reasons to vote against the construction of the dike were the costs because they would have been paid by the inhabitants. The further difference was *dikes* are purchasable and *dunes* were made by nature and have protected the assets and the people without any costs (Allemeyer, 2007, p. 93).

A comprehensive treatise of the knowledge on natural phenomenon and science of that time was prepared by Brahms (1754). Herein, Brahms described thoroughly the expertise on hydrology, hydraulics and civil engineering. He reduced the storm surge events and their consequences on the perceptions of natural sciences. This opinion might be strongly influenced by the enlightenment of that time. According to Allemeyer (2007) could the treatise of Brahms (1754) be taken as an example that the question was transformed from *if* to *how* human kind should interfere in natural processes?

In parallel, the opposite position was raised by a clergyman – see Allemeyer (2007). He has principal doubts that human kind should interfere in natural processes. One argument was that all efforts for protecting people and assets by dikes and the drainage of the hinterland derived their necessity from the given situation of that time, that is, settlement in a highly vulnerable coastal area. But he put this argument in a broader context whilst he generally asked if it would not have been wiser to stay with the traditional way of live. Building dikes was an interference of men into natural processes which will cause continuous work over centuries.

This description shows that already in ancient times the debate about the right way of protecting people against storm surges did occur. More details and literature can be found in Allemeyer (2006) and in Allemeyer (2007). It was the intention to highlight the fact that, looking back to past times, the development to the current status of coastal protection was not smooth and straightforward. On the other hand, current strategies differ not that much from the way of thinking of some people around 200 years ago. The main difference may be that today there is more and detailed knowledge available on natural processes. Furthermore, technical development extended and improved the possibilities of protecting people and assets against rising water levels.

5.1.2 Facts and Figures on Current Circumstances

Beforehand some facts will be shown which have been collated and published in recent years. The intention is to spotlight the (current) situation, although it is under continuous change. In Fig. 5.2 the flood-prone area[1] at the coast of the southern North Sea Region is shown. The elevation along the coastal stretches of the UK, The Netherlands, Germany and Denmark has been processed to visualize which area could be flooded if no protection elements (e.g., dikes or barriers) would have been installed. Additional to this map in Fig. 5.3 some figures have been collated to highlight how many people are living in the area and how big the extension of the flood-prone area is per country.

In these countries the 5 m contour line is the main inland border based on the highest (anticipated) water level for the extension of flooding, depending on specific circumstances a higher contour line might be applicable. The real extension of an inundated area in case of a dike failure would depend on the topography and on various other pre-conditions. For example, two-thirds of the Dutch area is flood prone, and many areas which are up to 7 m below sea level have been reclaimed from the sea by high efforts over the past centuries. In Lower Saxony, Northwest Germany, approximately 14 % are prone to floods, if no measures would have been taken. Storm surges in the Middle Ages extend the flood-prone area far into the inland since men fought against the rising water. Today approximately 6000 km^2 is protected against storm surges by a dike length of approximately 1200 km in total.

5.1.3 Historical Development

The first settlers (we don't care why they decided to settle in an inhospitable and vulnerable area) built their houses whether on the ground or on dwelling mounds – see, for example, Hallewas (1984); Reinhardt (1984) and van de Ven (1993). The reason why it was possible to live on the ground in that time without any protective elements may differ, but it is assumed that the regression of sea water enabled settling. With the experiences

[1] *Flood-prone area*: The area at the coast which will be inundated if no infrastructure against flooding are in place, for example, dikes or sluices. The extension of the flood-prone area differs from Member State to Member State. For example, in Germany the flood-prone area is bordered by the 5 m elevation line in the hinterland. At some coastal strips, the 10 m elevation line is the restriction, it depends on the calculated water level the dikes should withstand.

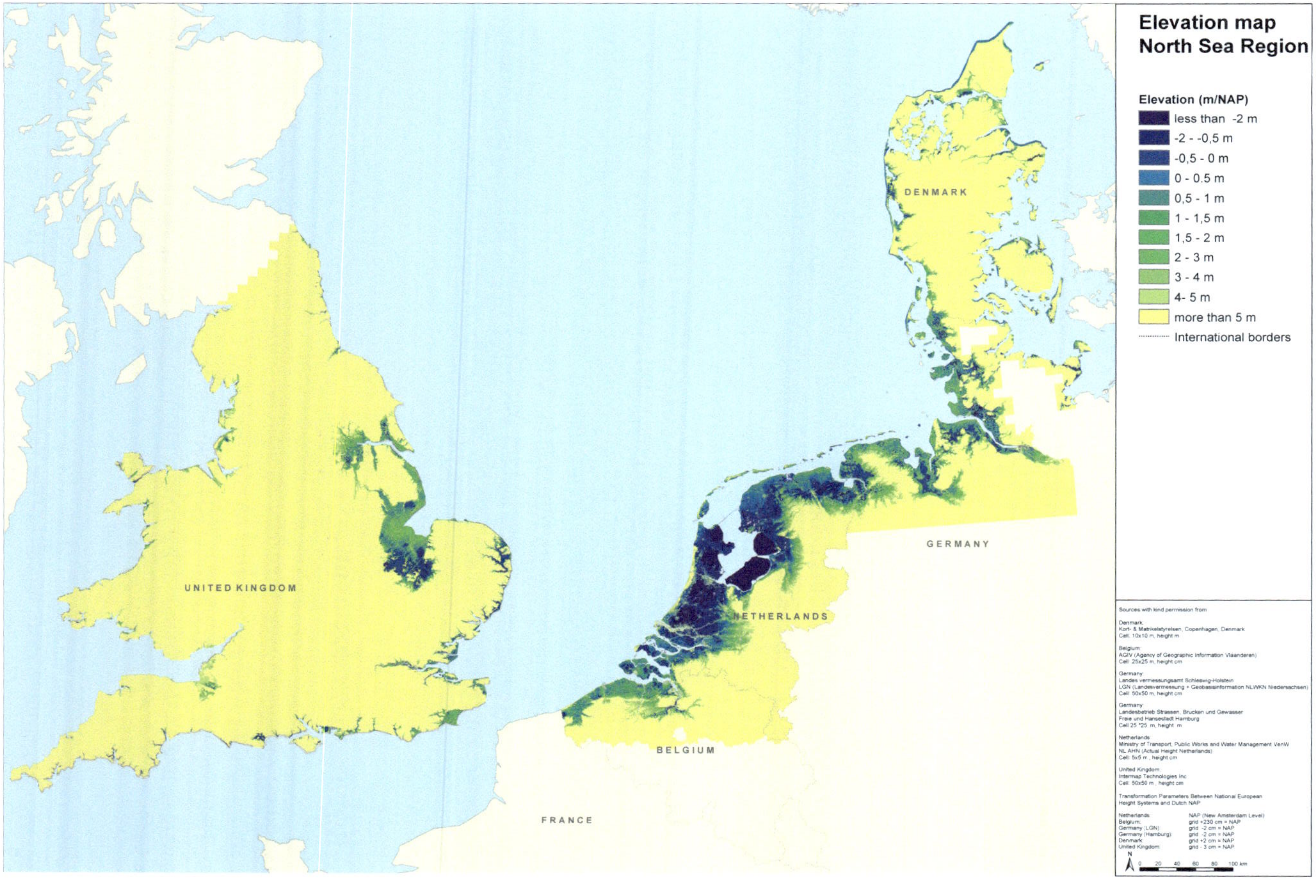

Fig. 5.2 The flood-prone area at the coast of the southern North Sea area (*yellow*: above +10 m, *green*: +5 to +10 m above sea level, *light* to *dark blue*: +5 m down to –7 m below sea level). Source: SafeCoast (2008, Annex 3.1)

	Denmark	Germany	Netherlands	Belgium	United Kingdom	*Round off totals*
Total coastline length[a]	4,605 km	3,524 km	1,276 km	98 km	17,381 km	*~27,000 km*
Developed 1 km coastal strip[b]	9%	11%	12%	48%	-	-
Area below +5 m sea level[c]	1,500 km^2	9,000 km^2	19,000 km^2	2,500 km^2	6,500 km^2	*~38,000 km^2*
Population below mean sea level[d]	< 5,000	1,800,000	9,000,000	380,000	2,500,000	*~14 Million*

Fig. 5.3 Summary of characteristic figures for the coast of the southern North Sea Region: (a) coastal length according to Doody et al. (2004), (b) figures according to EEA (2006, p. 11), within this report it is differentiated between the *immediate coastal strip (up to 1 km)* and the *coastal hinterland (between 1 and 10 km)* and the *non-coastal national territory*, (c) several national sources with varying figures. The term *area below +5 m mean sea level* is vulnerable against inundation if no measures against flooding would have been taken. Source: SafeCoast (2008, p. 19)

of higher water and the increasing and the enlargement of agricultural use of the fertile coastal areas improved protection was necessary. First, by building on higher land and afterwards by constructing low dikes. Investigations have been made by, for example, Blok (1984); Hallewas (1984); Reinhardt (1984); Brandt (1992); van de Ven (1993) and more recently by Knotterus (2005); Meier (2006); Behre (2008) and Meier (2014) to prove this development.

The *transitions of coastal protection strategies* is visualized in Fig. 5.4. As mentioned earlier first settlers adapted themselves by moving with rising and falling water levels.

Very likely, the next step was the accumulation of material to heighten the farm house. Many remnants (dwelling mounds) testify this theory; some of them can still be found in the coastal area of northern Germany, The Netherlands or southern Denmark (Fig. 5.5). These mounds were heightened after the experience of a flood or, perhaps, if the residents perceived an increasing water level. Also, the construction of a surrounding dike line of a farm house was as high as the lowest floods might came. Due to limited capability these dikes were approximately 2 m above ground level. The adaptation of these dikes was also based on the experiences of floods and the overtopping of waves during the storm surge season. Especially, in the beginning of the Middle Ages several storm surges led to a continuous process of dike heightening and strengthening. For example, Brahms (1754) prepared a thorough compilation of early engineering knowledge and expertise. Followed by many publications which enhance and amplify the knowledge basis, it is sufficient to refer to a selection of older publications such as Auhagen (1896); Tenge (1898); von Gierke (1917); Tenge (1912); Vierlingh (1920); Wöbcken (1924, 1932) and Breuel (1954).

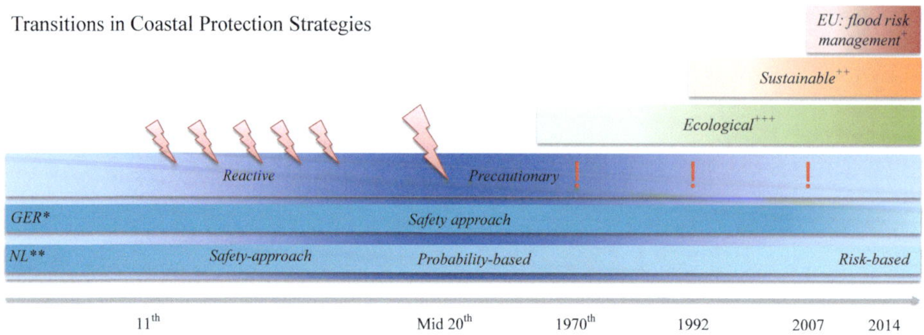

Fig. 5.4 Transitions of Coastal Protection Strategies in the southern North Sea Region. The flashes highlight the impacts of storm surge events. The exclamation points highlight the influence of the external triggers: ecological, sustainable and flood risk management. * Germany is a federal state; the national government is providing the legal framework and the implementation is made by the Federal States. Thus, the implementation strategies differ in each coastal Federal State. ** In The Netherlands a probability-based approach was developed and applied after the storm surge event in February 1953 (Recommendations of Delta Commission I). From 1990 to 2005 the probability-based approach was continuously enhanced and improved. Since 2014 a flood risk management approach is applied; see MinInfraM and MinEZ (2015)

Fig. 5.5 Picture of a dwelling mound in southern part of Jutland, Denmark. ©Frank Ahlhorn

In the nineteenth and twentieth centuries the strategies to protect people against storm surges differed, of course, from country to country, but at the end led to a closed dike line. The dikes were and are continuously improved by strengthening and heightening to estimated heights based on water levels of former storm surges (e.g., determined by the height of flotsam[2] on the outer dike slope). In the middle of the twentieth century this *reactive* approach changed into a more one *precautionary* (see, e.g., Vergouwe, 2015). Later on, this will be exemplified by the Dutch and the German approaches. Especially, in The Netherlands and Germany after the experiences of storm surge events with disastrous

[2]*Flotsam*: consisting of remainder of the vegetation from out the salt marshes and litter deriving from anthropogenic utilization of the sea and the coast.

consequences and many victims: After the severe flooding in 1953 in The Netherlands the approach to protect people and assets changed to a proactive work of civil engineers who calculated and constructed protective elements – see, for example, de Ronde et al. (1995) and TAW (1998, 2000).

The increasing economic wealth mainly based on agricultural production and trade led to the extension of the protected area. Longer dikes did not protect only a farm house but larger areas being used for cattle and sheep as well as peat and different types of crops. As the clergyman anticipated (see p. 64) building of dikes was accompanied by the necessity of drainage of the low-lying coastal area (Fig. 5.12 in Sect. 5.3). Furthermore, the ability to construct higher and stronger dikes led to a long dike line along the coast of the southern North Sea. Again, this was not a continuous and smooth story (see, e.g., Knotterus, 2005). Many storm surges, especially in the Middle Ages, led to many setbacks and a lot of victims. A list of these events for The Netherlands can be found in, for example, van de Ven (1993); Vergouwe (2015), for Germany, for example, Gönnert et al. (2001, Chapter 7.3.1).

The main question always was how high do we have to build the dikes to stand the next storm surges? At the same time it was only possible to determine the height of the water level by landmarks or the height of flotsam on the outer slope of the dike. But it was not possible to forecast the water level of the next storm surge. Consequently, this period can be called ***reactive***: the coastal residents mainly *re-act*ed to a disastrous event. Where possible, the ambition of the respective sovereign was to expand his area of influence. This could be done by an act of war or by reclaiming land. Land reclamation was only possible under certain conditions and was accompanied by a high effort in material, funding and manpower. The intention for land reclamation was solely the enlargement of the area of influence, that is, to obtain new and fertile agricultural land for the peasants, and not directly linked to the protection of people.

Brahms (1754) was one of the first who proposed a new way of thinking in that he pinpointed that the crest of the dike had to be higher than the former water level measured and that the outer slope of the dike should be more gentle than in former times. These conclusions were drawn from his comprehensive investigations on dikes along the Frisian coast at that time. Thus, Brahms (1754) already introduced a precautionary approach.[3]

The fading impression of disastrous events and the trust in strong dikes, which was proven without damages in many storm surges, led to the assumption that these dikes are save. During the first part of the twentieth century the work on protective elements was neglected or reduced to minimum. Especially, in the time of World War II, other engineering achievements were prioritized – see Blackbourn (2006).

Going back to Fig. 5.4 the first part from *reactive* to *precautionary* has been briefly delineated. The transition between these two strategies neither was smooth nor clear cut,

[3]Likely, the main actor, Hauke Haien, of Theodor Storms' novel "Der Schimmelreiter" was based on the person of Albert Brahms. Local residents at the coast casted aspersions to Hauke Haien because he wants to improve the profile of the dike by flattening the outer slope. The intention was to reduce the power of the waves which impinge on the clay surface.

but the publication of Brahms (1754) might be seen as one point of change, especially in Germany. The transition went differently in The Netherlands and in northern Germany, but was triggered by storm surges in 1953 (The Netherlands and the UK) and 1962 (Germany). Although the strategies have been revised, the way of doing so was different. Nevertheless, both approaches led to a similar result: the best possible protection of people and their assets against flooding at the coast. In general, the dikes were heightened and strengthened by the available techniques based on sophisticated models and methods to determine the height of future water levels.

5.1.4 Delineation of the Dutch Approach

The severe impacts of the storm surge in winter 1953 were a shock for all Dutch inhabitants. This event also caused severe impacts in the UK, for example, DCCF (1955),[4] but in The Netherlands approximately 1900 people lost their lives. Several dike failures led to enormous floodings – see, for example, Slager (1992); TAW (1998) and Seijffert (2001). A detailed description on how the safety standards were found and how the different Delta Commissions have worked in The Netherlands (Delta Commission I (1953–1960), Delta Commission II (1975–1977), Committee Boertien I (1992), Committee Boertien II (1993) and Delta Commission 2008) can be found in, for example, Yska (2009) and Voorendt (2015).

As a reaction to the disaster in 1953 the coastal protection strategy was totally reworked for The Netherlands (TAW, 1998). A master plan to increase the safety of the low-lying areas was established in the 1960s. The basis was the Flood Defense Act (FDA) established in 1953. Up until the 1950s the risk of flooding was estimated on the basis of intuition and experiences, as (complex) simulations or calculations to determine the risk of flooding were not possible. The first step of the advisory board of the Delta Commission was to investigate whether the water level of the 1953 flood (NAP[5] +3.85 m, Hoek van Holland) could be exceeded and the second step was to investigate the costs of increasing the safety level in comparison with the expected economic benefits (e.g., van Danzig, 1956; TAW, 1998, 2000). Based on the consideration of several uncertainties for the design of protective elements the Delta Commission I decided for NAP +5 m in Hoek van Holland as starting point for safety standards – see, for example, Peerbolte (1993). The result was the classification of the Dutch coastline and the rivers into four safety levels (see Fig. 5.6),[6] that is, ranging from 1 to 10,000 for highly populated areas with high economic values to 1 to 4000 at the Westerschelde estuary and the north-east of The Netherlands, and 1 to 1250 at the rivers – see, for example, Delta Commission (1961) and TAW (2000). The proposed coastal works, called "Delta Works", were largely completed in the late 1980s.

[4]Further information: www.metoffice.gov.uk/news/in-depth/1953-east-coast-flood

[5]Normaal Amsterdam Peil = Amsterdam Ordnance Datum.

[6]The safety levels are exemplified by the risk of failure of the protection infrastructure by over topping.

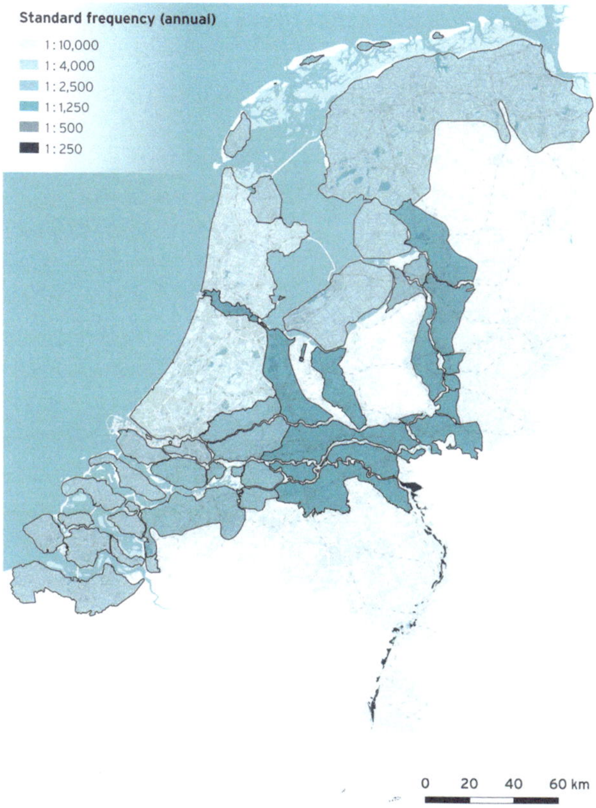

Fig. 5.6 The map shows the currently defined safety standards laid down in the Water Act of 2009 (MVenW, 2010) in The Netherlands. Source: Vergouwe (2015, p. 12)

The approach was based on the suggestions to strengthen the embankments as a result of the experiences of the storm surge in 1953 and the goal was to retain the water from the hinterland.

This probability of exceedance-based approach has been applied to the entire coast of The Netherlands. From 1953 to the late 1980s many dikes, sluices and storm surge barriers have been improved – and to some extent newly installed – to increase safety against flooding. However, the design of the embankments was subject to the lack of experience and knowledge of the various failure mechanisms, which could lead to a dike failure (TAW, 2000). The Technical Advisory Committee for Flood Defence published comprehensive guidelines for the maintenance and the construction of embankments and the consideration of different values and functions in the coastal zone and along the rivers – see TAW (1998). Shortly after the completion of the Delta Works a new policy of dynamic maintenance was established, accompanied with the baseline concept for the sandy coast (MVenW 1990; Hillen and de Haan 1993). It was a reaction to the enduring coastal erosion along the coasts of The Netherlands and the anticipation of possible consequences of an accelerated sea level rise. Applying the baseline concept erosion can be allowed to take place to a limited extent, but will be prevented on highly vulnerable coastal stretches. The main technique deployed to implement this strategy was the application of sand nourishment and this has led to good results. The concept of the coastal defense strip was introduced from the –20 m depth-line on the seaside to the polder on the landside (MVenW, 1990).

The Netherlands applied a scenario-driven approach for an accelerated sea level rise, which was based on the IPCC scenarios and ranges from 20 cm/100 years (minimum scenario) to 85 cm/100 years (maximum scenario) with a 10 % increase in wind speed (MVenW, 2002). The safety level of the embankments will be reassessed every 6 years.

The Technical Advisory Committee has stimulated an investigation of the failure mechanisms of coastal protection elements. Consequently, a large research project has been launched called VNK (Veiligheid Nederland in Kaart, Flood Risks and Safety in The Netherlands, Floris). The approach taken in this project was described as: “From probability of exceedance to probability of flooding” (TAW 2000, MVenW 2005a-c). This means that the safety margins established by the Delta Commission are still valid, but the design water level as criteria for the dike height will be substituted by an acceptable risk.

The definitions are according to the Committee as follows (TAW, 2000, p. 8):

Exceedance frequency: The exceedance frequency of a water level is the probability that the design water level is reached or exceeded. The design water level is used to design a safe dike or hydraulic structure.

Flood probability: The flood probability is the probability, that an area might be inundated because the water defense around that area (i.e., a dike ring) fails at one or more locations.The failure of a protective element can have different causes.

The background of the conceptual shifting in The Netherlands refers to the fact that about 50 dike rings around polders have been installed to protect people against flooding and at the rivers against inundation. The safety of the dike ring is provided by different constructions such as dikes, sluices and locks. Every element has its own failure mechanism and its specific failure probability, especially, that of human error, as the experiences of 1953 have shown (Slager, 1992). The aim of the Floris project was to determine as exactly as possible the probability of failure of these components. This was done by assuming that a chain is as strong as its weakest link. Thus, the Floris project identified *Weak Spots*. This approach and the paradigm shift has been integrated as early as possible in a much wider approach to coastal development – see MVenW (2002). The final result of the second phase of the Floris project was published in Vergouwe (2015) and will be elaborated in Sect. 6.1.4.

5.2 Delineation of the Development in Northern Germany

There are four coastal states in northern Germany responsible for coastal protection. This summary should be seen as an example how the responsible institutions intended to tackle the challenge of protecting people and economic values against storm surges.

As a reaction to the storm surge of 1953 in The Netherlands an Engineering Commission was established in northern Germany to investigate and to improve the existing strategy of coastal protection – see, for example, Lorenzen (1955) and Tomczak (1955). The recommendations led to higher crests on some coastal stretches which later withstood the water heights and the wave run-up of the 1962s storm surge. The storm of 1962 led to extreme water levels in the area of the rivers Elbe and Weser, especially in Hamburg and Bremen. The "Lower Saxonian program for coastal protection 1955-1964", installed after the Dutch disaster, was not completely implemented by February 1962. The water levels at the East Frisian coast were less than expected; about two-thirds of which were not damaged, but the consequence of 61 dike breaches was an inundated area of approximately 37,000 ha (Engineering Committee 1962a). The Engineering Commission acknowledged that older dike lines, which had been built with a different mission (i.e., to protect the polder against inland waters from the peaty area), had protected the hinterland against devastating inundations. Thus, the recommendation was given to maintain and improve the older dike lines, and where possible to build new second dike lines *(= additional levees in the hinterland)*. Furthermore, it was recommended that the second dike line should be installed within a certain distance of the main dike to enable better emergency management in the case of a storm surge and a dike failure. The experiences in 1962 have shown that the response time in some cases was too short to rescue the inhabitants of the polder areas (Engineering Committee, 1962a).

The experiences of the 1962 disaster led to the installation of the Advisory Committee for Coastal Protection for the North and the Baltic Sea (Lorenzen, 1966). The comprehensive recommendations of this Committee were published in Engineering Committee (1962b). These recommendations cover the design of dikes, the design of the outer and inner slope, the quality and conditions of the soil and the subsoil of dikes, and the maintenance and contingency planning. The results of the research programs and the recommendations of the committee led to the first ever State Law on Dikes in Lower Saxony – see NDG (1963).

The coastal protection authority is obliged to prepare a special plan describing and elaborating the strategy and concepts of coastal protection (Master Plan for Coastal Protection – German: Generalplan Küstenschutz). The first Master Plan, published in 1973, included a comprehensive description of the consequences of the storm surge in 1962, objectives of coastal and island protection in Lower Saxony and the design of main dikes and their heights (Fig. 5.7), and also the work program for the coming years – see Nds. MELF (1973).

Fig. 5.7 The shape of a main dike in Lower Saxony, northern Germany. The photograph shows the sea side of the main dike with a gentle slope and a salt marsh in front of the dike. ©Frank Ahlhorn

Currently, coastal protection in Germany is based on the Constitution of the Federal Republic of Germany (German: Grundgesetz). In article 74 section 1 coastal protection is an integral part of the concurrent legislation between the Federation and the Federal States (Deutscher Bundestag, 2007b). In practice, the responsibility lies with the Federal States (principle of subsidiarity). All coastal states have a special legislation for water management and all of them have incorporated coastal protection into this framework with the exception of Lower Saxony with its State Law on Dikes. The Federation contributes financial support to this task, established in article 91a of the Constitution (Deutscher Bundestag, 2007b). The contribution is established as "Federal Objection for the Improvement of the Agrarian Structure and Coastal Protection". The aim for coastal protection is as follows: "Defence against natural hazards and enhancement of safety in coastal zones, on the islands and on river basins in tidal areas against inundation and loss of land by storm surges and sea attack" (Deutscher Bundestag, 2007a, p. 62). Other provisions are included, such as strategy development and investigations in combination with measures, to reconstruct, strengthen and heighten coastal protection structures, for land work within a range of 400 m and sand nourishment. Limited support is given for coastal protection measures, which affect areas of ecological value. These are only eligible if, for example, the required safety cannot be achieved by another justifiable measure. Measures which are not covered by the Federal Objection are, for example, the maintenance of coastal protection structures and the construction of pumping stations.

The recommendations of the Engineering Committee for Coastal Protection of the North and Baltic Sea are still valid (Engineering Committee, 1962b) and have been amended in the light of recent findings in 1993 (EAK, 1993) and 2002 (EAK, 2002). The latest Master Plan for Coastal Protection in Lower Saxony in combination with Bremen (Mainland) was published in 2007 (NLWKN, 2007). The concept and the strategy of coastal protection has not been changed since the Master Plans of 1973 and 1997.

The methodology for calculating the height of the dike is recommended by the Advisory Committee for the North and Baltic Sea and combines two approaches: (a) Single Value Procedure and (b) Composition Procedure (Fig. 5.8).

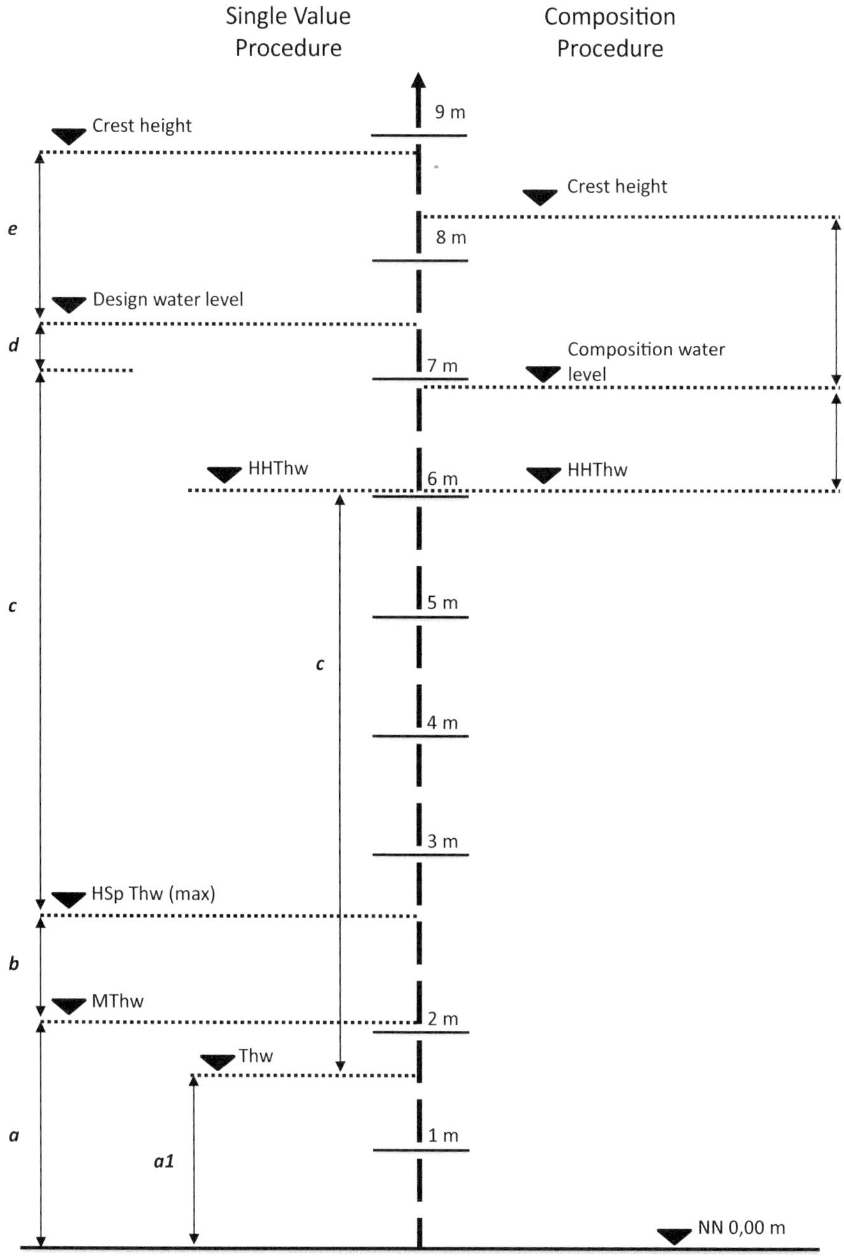

Fig. 5.8 Procedures to calculate the design water level and the crest height of the main dike, left: single value, right: composition. *Thw*: tidal high water, *MThw*: mean tidal high water, *HHThw*: highest tidal high water, *HSpThw*: highest spring tidal high water; *a*: mean tidal high water, *a1*: calculated astronomical tidal high water including wind effects, *b*: maximum increase by tidal spring water, *c*: highest water level raised by wind (highest high water level – a1), *d*: safety margin (including regional sea level rise, soil subsidence, etc.), *e*: maximum wave run-up, *: regional sea level rise. Source: adapted from Bezirksregierung Weser-Ems (1997)

The aims of safety for coastal protection at the North and Baltic Sea are summarized as follows:

The basic findings of the Advisory Committee for Coastal Protection – see EAK (2002) – are as follows:

- The absolutely highest storm surge can not be determined,
- certain specific astronomical, meteorological and oceanographic circumstances coincide may leading to higher storm surges as anticipated today,
- it is questionable whether dikes should be built which will hold back flooding water under all perceivable conditions.

This led the Advisory Committee to the following recommendations for a main dike (Kunz, 2004b, p. 256–257):

- The height of the dike is crucial for the safety of the protected marshes,
- the crest of the dike should be as high as the design water level, which is based on a defined level of safety taking into account a certain probability of exceedance and the height of wave run-up, which should prevent frequent and powerful over topping,
- the decision not to determine a "highest storm surge", has to be considered by designing dikes with a certain cross-section able to withstand strong and long-lasting over topping.

For the design of main dikes the Committee recommends

- For every stretch of the coast the "design water level" has to be determined – verified by scientific research and clearly defined – which either will never or at least within a manageable risk level – be exceeded within a certain time period. The currently applied procedures (in Schleswig-Holstein with regard to the frequency of a certain water level, and in Lower Saxony to the single value procedure) produce approximately the same values, but are not verified by science and are more or less the result of empirical studies,
- identification of a "design height for wave run-up", which will only be over topped by a certain wave level or only allow a defined amount of water to flow over the dike crest (overflow).

These guidelines of the Advisory Committee are reflected in the current procedure to determine the height of the dike crest and, thus, in the accepted safety standard, incorporated in the procedure applied in Lower Saxony as shown in Fig. 5.8. Since no absolute safety against flooding can be guaranteed, the risks have to be limited and managed. There are many uncertainties which have to be considered to determine the safety level of embankments. As mentioned above, for example, the highest storm surge cannot exactly be determined, for example, because of uncertainties in projecting water levels into the future. There is no doubt that the existing safety concept has to be enhanced; however, step by step and not immediately by introducing new approaches before they are ready for implementation, for example, it is necessary to enhance the knowledge of failure mechanisms and the failure probability of dikes and other structures.

Taking these remarks into account, the recent enhancement of the safety margin for the secular sea level rise from 25 cm up to 50 cm announced in 2007 by the Minister of the Environment, responsible for coastal protection, is a postponement of the issue at stake. However, it serves as an approach to address the consequence of increased sea level rise in Lower Saxony (and Bremen).

The development of different procedures on how to determine the crest height of a dike is a tribute to the problem of calculating this height. An attempt to reduce or to minimize the uncertainty of the extend of a worse event. Some of the parameters shown in Fig. 5.8 contain uncertainties or at least the circumstances or conditions to measure them are under discussion. For example, the value of the mean high water level is a calculated value based on a certain time period of high water levels. The parameter highest high water level differs from coastal stretch to coastal stretch. This parameter is influenced by several circumstances which also contain uncertainties. The parameter *e* maximum wave run-up also depends on global conditions (e.g., wave height in the North Sea, wind speed, fetch) and local conditions (e.g., degree of shelter by islands or tidal flats, orientation of tidal channels in front of the dike line). Consequently, the procedure to calculate the crest height shows the best available approximation and is a reliable way to be dealt with by a government. Nevertheless, engineers are continuously working on the improvement.

Another consequence is that if the crest height of a dike is calculated based on this procedure everywhere in the Federal State the safety level is the same. That means according to the risk approach (see Sect. 6.1.2): is every area is protected by the same safety level the (residual) risk differs. For example, some areas are densely populated and industrial areas are protected by dikes. Other areas are dominated by farm land and are less populated. The consequences of a dike failure would vary between both areas. The densely populated area is more vulnerable than the farm land, but both are protected by the same safety level. That means, even if the safety level stays the same risk differs along the coast depending on the values and the equipment of the area of concern.

Going back to Fig. 5.4 in the 1970s another external trigger imposes the consideration of ecological aspects within coastal protection. This is exemplified by Ley Bay case (East Frisia, Lower Saxony, see in Fig. 8.3 within the description of problem 3 in the exercises in Sect. 8.1):

During the 1960s three different reasons demanded for an improvement of the drainage system and coastal protection around the Ley Bay (Fig. 5.9). Caused by storm surges the Ley Bay was enlarged approximately 40 km inland. Continuous land reclamation work reduced the dimension and generated new fertile land for peasants until the beginning of the twentieth century. The process of sedimentation continued and after World War II and new land should be offered to the arriving refugees. Therefore, it was planned to reclaim a large amount of new land at the Ley Bay. In the middle of the twentieth century a new dike was built to protect the new land against flooding. Afterwards, the sedimentation process did not stop in the Ley Bay due to different reasons. The circumstances were getting

Fig. 5.9 Tidal creek in the Ley Bay. ©Frank Ahlhorn

worser for fishermen to reach the open North Sea from harbors around the Bay. Drainage of the hinterland was limited due to the siltation of tidal channels in front of the tidal gates and the main dike had to be improved. These reasons led to the idea to close the Bay in order to solve all problems and to offer new attractive features for tourism. At that time ideas of nature conservation emerged, namely the protection of areas for birds. The consequence of closing the Bay would have led to the transformation of a saline environment to a fresh water environment. Models for this transformation existed in The Netherlands, for example, the closure of the Lake Ijssel in the 1930s. Nevertheless, protest was raised and a court decided that the Ley Bay has an outstanding natural value, so an alternative solution has to be developed. In the middle of the 1990s the solution was finally implemented. This process could be seen as the first point where nature conservation influenced the conduction of coastal protection work. Figure 5.4 highlighted this event as ecological consideration in coastal protection strategies. Coastal protection management was urged to consider ecological aspects, because Lower Saxony installed a national park in front of the main dike in 1986. The same development could be observed in Schleswig-Holstein. Until then many attempts have been taken to consider more ecological aspects. For example, by joint development of a strategy for the dike fore land which is under the protection of the national park administration as salt marsh – see, for example, Dijkema (1987); Ovesen (1989); Stock et al. (1994); Hofstede and Schirmacher (1996); van Duin et al. (1999); NLWK Norden (2003) and NLWKN (2007). Furthermore, in recent times under specific conditions building of dikes has to be treated by a planning approval procedure. This procedure contains the conduction of Strategic Environmental Assessment (SEA) and Environmental Impact Assessment (EIA).

Exercise

1. Compare the coastal protection strategies of The Netherlands and northern Germany. Beforehand, select a Federal State of Germany because the strategies are slightly different. What are the similarities and what are the differences? More important, what are the similarities or differences according to *real* implementation, for example, for the construction of dikes?

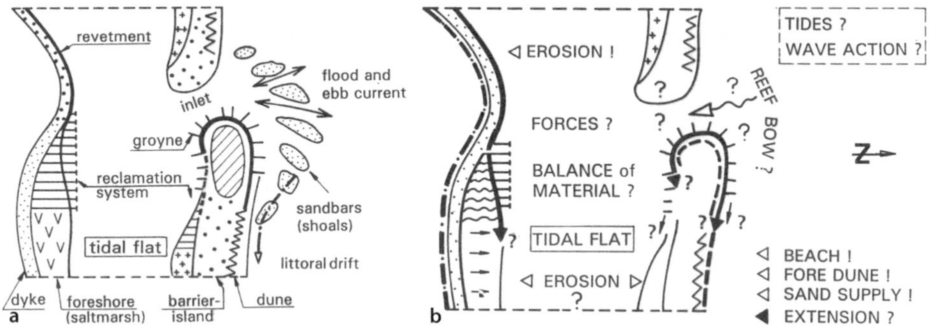

Fig. 5.10 (**a**) The cross section shows the natural elements of a coastal stretch with (barrier) islands. From the main land (*left*) over the tidal flats to the (barrier) islands (*right*): salt marsh, sand and mud flats (tidal flats), dunes, sandbars, inlet. Different artificial elements are mentioned: revetment at the main land, groynes, reclamation systems. (**b**) What will happen in future time with the coastal system under the influence of climate change? Source: adapted from Kunz (2004a, p. 98)

2. Find further examples for the influence of ecological aspects in coastal protection in other Member States in the North Sea Region.
3. Coastal protection has not only to safeguard the main land coast, also islands have to be protected against waves and currents, and, of course, against storm surges. Some islands consist of clay *and* sand. Dunes are protecting the main part of the islands against erosion and wash overs, and, as it has been expressed in 1650, the islands also protect the main land (see the description from Bartels (1881) on p. 63). Figure 5.10 shows the current situation (a) of a coastal stretch as it can be found in the southern North Sea Region. What might happen in future time to the system of (barrier) islands and main land? Describe the interaction of this system.

5.3 Drainage Management

5.3.1 Delineation of Historical Development

The first settlers were urged to protect themselves against high water either from the sea or from the hinterland. The landscape drawn in Fig. 1.3 provides an impression how the coastal area looked like in ancient times. Tidal channels were not cut off from the land, there was a gradual transition from the flood plain to the tidal area. Higher areas contained forests, lower areas were populated by salt-resistant vegetation. The entire coastal system was in a dynamic equilibrium depending on hydrodynamic conditions (see Chap. 1).

Evidence for the oscillation of the high water line (or the calculated mean sea level) at the coast are given, for example, by Jelgersma (1979); van de Plassche and Roep (1989); Streif (1989, 1993); Denys and Baeteman (1995); Behre (2003, 2007) and Baeteman et al.

(2011). The question on the accuracy and correctness of the sea level curve (showing the development of the sea level in a certain area – see, e.g., Behre (2007); Baeteman et al. (2011)) is in our context insignificant, but has to be considered for the interpretations of archaeological finds.

The first settlers started with building of dwelling mounds. Furthermore, the arable land was enlarged and small dikes were constructed. Figure 5.11 gives an impression of a situation when the main dike line was protecting a huge area against flooding. In ancient times people produced salt from frequently flooded peat areas by digging and burning (the process is called *selnering*) caused by an increasing demand of salt – see, for example, Marschalleck (1973); Besteman (1974); Fries-Knoblach (2001); Nienhuis (2008) and Siegmüller and Bungenstock (2010). Additionally, farm land for crops and cattle was enlarged and should be protected against fluvial and pluvial floods. Consequently, there was a strong need for efficient and swift drainage of these polder areas. Drainage management was additionally challenged by selnering because it led to increased subsidence of the surface. Some authors assume that this kind of land use entailed the beginning of land loss in the southern North Sea Region. Thus, innovation was required. The protection against storm surges, the enlargement of arable land and the production of salt by selnering demanded new water management options.

Normally, the surface water flows caused by the gravitation "downhill" and discharges in bigger and bigger rivers and, finally, into the sea or a lake. In coastal areas influenced by tides the system of drainage is much more complicated, because the water level has a strong influence on the possibility of drainage (see Chap. 4). Water bodies which are

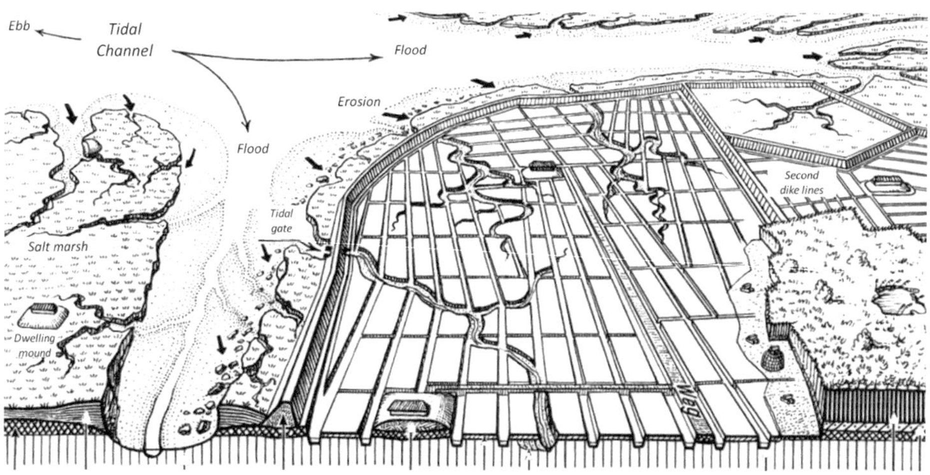

Fig. 5.11 Sketch of coastal landscape with dikes protecting the productive area and settlement. *Right*: Tidal channels have been closed off from the sea, drainage is organized by tidal gates. The excavation of salt (selnering) led to sinking land surface and, thus, to increase demand for inland drainage. To limit the flood prone area in case of a failure second dike lines have been constructed or are remnants of former main dikes in the hinterland. Farm houses were built on higher ground either on dwelling mounds or on the moraine. *Left*: The open landscape still exists, tidal channels were not closed and farm houses were built on dwelling mounds. Source: adapted from Bantelmann (1966)

open to the sea have changing water levels due to high and low tide. During high tide the (sea) water level is flowing upstream into the rivers or small tidal channels (Fig. 1.3). The construction of a closed dike line reduced or even prevented the free flow of the tides. Infrastructure has been installed to control the inflow and the outflow of the water into the protected areas. The first drainage means in the dike were wooden gates or tubes (Brandt, 1984). The wooden gates opened and closed automatically with the increasing or decreasing water level of the sea. Thus, a certain amount of time exists where the surface water can be discharged into the sea (in German: Sielzugzeit, in English: time period of discharge, see Fig. 4.12). This way of drainage is strongly depending on the topography, thus, it is only possible where the topography allows gravity driven discharge. If the topography is lower than the sea water level technical solutions have to be installed to actively pump the water into a canal or river (see Fig. 5.13).

The increasing demand led to the installation of wind mills, starting approximately in the fourteenth century (Fig. 5.12). Over centuries the advancement of traditional means influenced by the industrialization and by the implementation of pumping stations powered by combustion or electric engines made progress. Consequently, water boards did no longer depend on the wind to operate the wind mills. This offered a new option for water management, in wet periods the technical means allowed to drain the water preventive. Figure 5.12 exemplifies the development and interrelationship between drainage and subsidence of the hinterland for The Netherlands. Drainage of the land led to decreasing

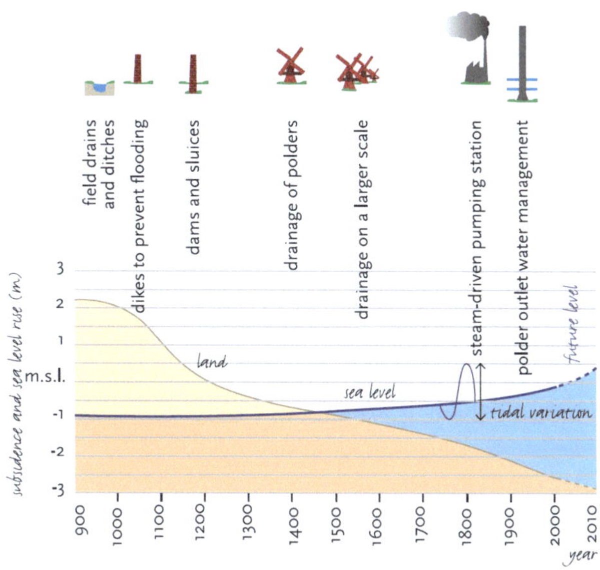

Fig. 5.12 Diagram showing the interrelationship between drainage by different means and the subsidence of land caused by drainage management. Exemplified for The Netherlands starting more than 1000 years ago with the drainage of low-lying areas and the construction of first dikes. Despite the fact that the surface is sinking sea level is rising due to different reasons, for example, climate change, which caused the same effect for drainage management: increasing height difference. Source: RWS (2011, p. 15)

surface which got even worser by the installation of technical means, such as wind mills and, of course, mechanical pumping stations.

Water management in Germany is organized by small self-acting associations, called water boards. Similar organizations are working in The Netherlands. In both countries today's water boards originated from small drainage boards responsible for the drainage and management of small areas. In the course of time small water boards were merged to bigger ones. For example, in Lower Saxony (north-western Germany) approximately 90 water boards are currently responsible for the water management (maintenance of infrastructure and water bodies); in The Netherlands there are 23 water boards active. In contrast to Germany the Dutch water boards have more power and responsibilities. In The Netherlands centuries ago there were approximately 1500 water boards existing.

The area of a water board is mainly coinciding with a certain catchment area of a discharge channel (either this channel is artificially created, thus, a canal, or an almost "natural" ditch).[7] In some cases the catchment area is divided into sub-catchment areas because of the topography of this distinct area. Then the surface water of this sub-catchment is discharged to sub-channel and, afterwards, pumped into the main drainage channel.

Due to the topographic conditions of low-lying land, approximately down to –4 m below sea level, several infrastructure are installed in the hinterland to control and to discharge the water. At some places the fresh water has to be pumped four times to surmount approximately 12 m of elevation.

The elevation in northern part of Lower Saxony (Germany) is ranging from –4 m below sea level up to +12 m above sea level. The low-lying areas are marsh land, former salt marshes and remnants of Middle Age storm surge disasters. To get a much clearer picture in Fig. 5.13 the crosssection from the highest point in the area (the moraine), down to the lowest points in the peaty area and up to the river Ems (bordering Germany and The Netherlands) is shown. The surface water flows through smaller and bigger ditches or canals "downhill" to the lowest parts of the areas (here, e.g., from +12 m down to –2 m). Through a system of ditches and canals the surface water is pumped into the river Ems to finally discharge the water into the North Sea.

Gradually, the surface of the area between the dike and the moraine is still shrinking or shrinking caused by

- extensive drainage of peat and marsh areas,
- soil compaction,
- soil excavation,
- mineralization,
- post glacial subsidence and
- extraction of fossil fuels from the underground.

[7] Almost all water bodies in low-lying coastal areas are artificial or at least heavily modified by men, see, for example, the classification of the water bodies according to the European Water Framework Directive (see Ch. 6).

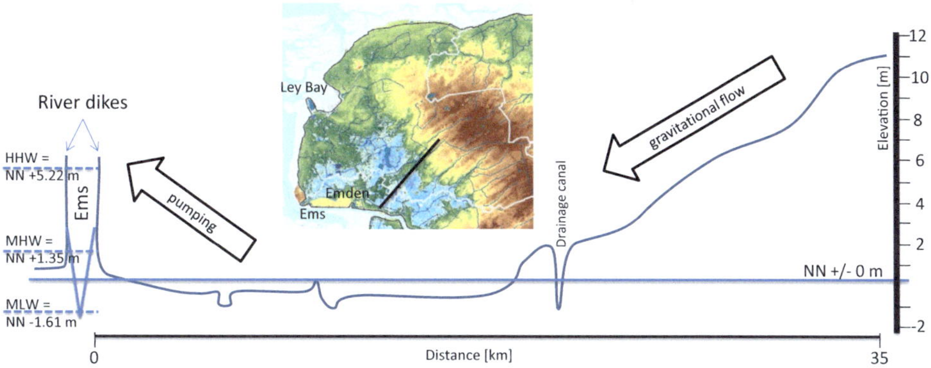

Fig. 5.13 Cross section from the moraine (*right*) down to the river Ems (*left*, north-west Lower Saxony, *MHW*: mean high water, *MLW*: mean low water, *HHW*: highest high water). Source: adapted from Kramer and Rohde (1992)

Typology of Drainage in Tide-Related Coastal Areas: A Suggestion

Different reasons are able to result in further subsidence of the surface and, thus, to an increasing challenge for drainage. Based on this description a typology is suggested which was first published by Kunz (1975) for hydrological investigations in tidal coastal areas:

Type A Areas of sufficient height are directly linked to the main discharge channel. A storage or tidal gate exists.

Type B The area could mainly be drained by gravitation. In case of high water levels (either inland or at the sea) a pumping station is installed.

Type C In this area the demanded water levels could not be achieved by gravitational discharge. The entire drainage is done by pumping stations.

Type D The area is completely drained by technical infrastructure. Partly, sub-catchment areas are also drained by artificial means, that is, sub-pumping stations.

The proposed typology for drainage management, especially in tidal areas, could be used to illustrate the sensitivity of this water management system against the impacts of climate change, that is, sea level rise (see exercise on p. 86).

5.3.2 Final Remark

The previous paragraphs concentrate on selected aspects (here mainly water quantity) of water management. Therefore, the following pages provide a selected collection of definitions and principles for water management. The intention is to briefly highlight that

Fig. 5.14 Drainage canal in the Wesermarsch (Lower Saxony), Germany. ©Frank Ahlhorn

much more duties and tasks are connected to water management. It is not possible and would go beyond the scope of this book to comprehensively summarize these different fields of water management (Fig. 5.14).

Definition in the USA

The legal definition of *water management* in the USA[8] is as follows:

> States and local governments establish water management departments to oversee the drainage of surface water and the reclamation of wetlands, swamplands, overflowed lands and tidal marshes and flood prevention and the conservation, development, utilization and disposal of water. Some of the duties of water management include:

- To locate and establish levees, drains or canals and to cause to be constructed, straightened, widened or deepened any ditch, drain or watercourse;
- To construct for the purposes of flood prevention or the conservation, development, utilization or disposal of water works of improvement, including levees, embankments, floodwater retarding structures, water storage structures, outlets and tide gates, flood gates and pumping plants for preventing floods, providing drainage, reducing sediment and reclaiming wet, swamp or overflowed lands and other related works of improvement that will carry out the purposes of this article; and
- To provide maintenance for such installations.

Definition in Australia

The second definition and primarily focusing on the principles of water management was taken from Australia.[9] Water management has mainly to deal with

---

[8]See: definitions.uslegal.com: access Apr. 2017.

[9]See: www.austlii.edu.au: access Apr. 2017.

- water sources, floodplains and dependent ecosystems (including groundwater and wetlands) should be protected and restored and, where possible, land should not be degraded, and
- habitats, animals and plants that benefit from water or are potentially affected by managed activities should be protected and (in the case of habitats) restored, and
- the water quality of all water sources should be protected and, wherever possible, enhanced, and
- the cumulative impacts of water management licenses and approvals and other activities on water sources and their dependent ecosystems, should be considered and minimized, and
- geographical and other features of indigenous significance should be protected, and
- geographical and other features of major cultural, heritage or spiritual significance should be protected, and
- the social and economic benefits to the community should be maximized, and
- the principles of adaptive management should be applied, which should be responsive to monitoring and improvements in understanding of ecological water requirements.

Further on, the Australian definition focuses on specific items with regard to water issues such as water sharing, water use, drainage and flood plain management. All of these items contain the pretension to avoid or to minimize the degradation of water bodies and to preserve or improve the functionality of adjacent marsh lands by anthropogenic measures.

As an example the intension on aquifer management is highlighted where "interference activities must avoid or minimize land degradation, including soil erosion, compaction, geomorphic instability, contamination, acidity, water-logging, decline of native vegetation or, where appropriate, salinity and, where possible, land must be rehabilitated, and the impacts of the carrying out of aquifer interference activities on other water users must be avoided or minimized".

Definition in Germany

In Germany the term *water management* is translated as *Wasserwirtschaft* and is defined in the German DIN standards (DIN 4049) as follows: "Water management is the goal-oriented order of all human interactions on surface and subsurface water bodies." According to §1a of the German Water Management law shall all water bodies be managed for the collective good and in consistent with the individual use, and that every avoidable interference shall be omitted. Since the German Environmental Program from 1971 the aim of water management is

- to preserve or re-establish the ecological equilibrium of the water bodies
- to ensure the water supply of the population and the economy, and
- in parallel, to enable all other utilizations of the water[10] for the common welfare in the long run.

[10]Explanation for *all other utilizations*: the use of water for leisure and recreation, for fishery, for energy production (to impound and to lower the water level), for the irrigation of agricultural areas and the discharge of sewage water.

To not burst the scope of this book the German definition will mainly be taken as orientation. This does not imply that other definitions are debased.

Exercise

1. Essential for water management either coastal protection management or drainage management is the sea level. Within this section uncertainties for the determination of sea level change in former times have been mentioned. Which uncertainties have to be considered if the relative water level is investigated based on, for example, archaeological finds? What are the differences regarding sea level change between the different coastal areas along the southern North Sea Region?
2. Sea level change is one of the main driving factors in coastal areas. Especially, in low-lying coastal areas sea level is an important constraint for coastal protection as well as water management. There are different terms linked to sea level change, the following questions are related to these terms. What is meant by
   a. Transgression
   b. Regression
   c. Eustatic change of sea level
   d. Isostatic change of sea level
   e. Which *type* of sea level change is linked to global, which one to local changes?
3. In Fig. 5.12 the subsidence of the surface is shown and exemplified for The Netherlands. Has this process stopped or will it stop in the future? What are the problems and/or challenges in the future if subsidence continues? Which options can be taken by water management to counteract this process?
4. In Sect. 5.3.2 an excerpt of the understanding of water management tasks and duties has been provided. The previously given historical delineation explains the development from the viewpoint of necessities emerging by the utilization of coastal lowlands by settlers and farmers. If you compare the needs and interests of the historical development with the current tasks of water management and the anticipated future changes what are the main challenges for water management?
5. The concentration of this section is on water quantity, the problems linked to water quality haven't been touched so far. Try to find out which interactions and interferences between different types of land use and water quality are due in coastal areas.

5.4 Estuaries in the Southern North Sea Region

Estuaries indicate the transition between open sea and inland waters and the gradual transition between land and water, by, for example, shallow water areas and periodically inundated embankments (see Fig. 1.3). At estuaries the transportation hubs for worldwide trade are located and specific habitats can be found. Thus, they are also a focus of integrated coastal zone management.

This section provides a brief outline about selected estuaries in the southern North Sea Region: Elbe and Weser (both in Germany) and Schelde (estuary crossing the borders of Belgium and The Netherlands). The descriptions in this section represent the situation and development of many estuaries around the world – see, for example, Gregory (2006); Ducrotoy (2010) and Collins and Miller (2011).

5.4.1 Introduction

An estuary is a transition zone between a river and the open sea; fresh water is mixed with salt water. Rivers in the North Sea are influenced by semi-diurnal tides. The consequence is that flora and fauna have to adapt to these changing circumstances. Specialized plants and animals can be found which are adapted to saline or to fresh water environment and some of them are able to cope with both conditions.

Due to the fact that the transition from fresh to salt water is one of the main characteristics of estuaries in 1958 the Venice system (Anonymous, 1958) was developed to differentiate between water bodies based on salt concentration. Attendees of an international Symposium for the Classification of Brackish Waters in 1958 agreed on four categories: hyperhaline ($\geq \pm - 40\,\%$ salinity), euhaline ($\pm 40 - \pm - 30\,\%$ salinity), mixohaline ($\pm - 30 - \pm - 0.5\,\%$ salinity) and limnic ($< \pm\, 0.5\,\%$ salinity). The $\pm$ indicates that these figures should not be taken as precise limits, more as approximation.

The mixture of fresh and salt water is called brackish water body. The brackish water body provides no strict border between the saline and fresh water environment. Mainly driven by tides the brackish water body is oscillating in a certain reach of an estuary (see Chap. 4). It should be understood as a brackish water zone as transition between fresh and salt water. The extension of the brackish water zone depends on many circumstances, for example, fresh water discharge from upstream the river, the tides, the discharge of tributaries and of the shape of the estuary. If there is low fresh water discharge the sea water can penetrate further upstream (Fig. 4.9).

Caused by tidal range there is also a transition between dry and wet areas. Shallow water areas and temporarily inundated sand and mud flats exist in an estuary with endemic flora and fauna. Also anabranches and side arms are important ecological areas in estuaries.

Figures 5.15–5.17 provide an impression that the three estuaries show differences according to the classification of the Venice system. Depending on the orientation to the sea, human interferences and the length of the distinct zones show different extensions.

The Weser estuary has a total length of approximately 125 km and according to morphological purposes it is divided into two parts: the Lower Weser from Bremen to Bremerhaven (70 km) and the Outer Weser which extends from the harbor of Bremerhaven to the North Sea (55 km). According to the Venice system the Weser is divided into

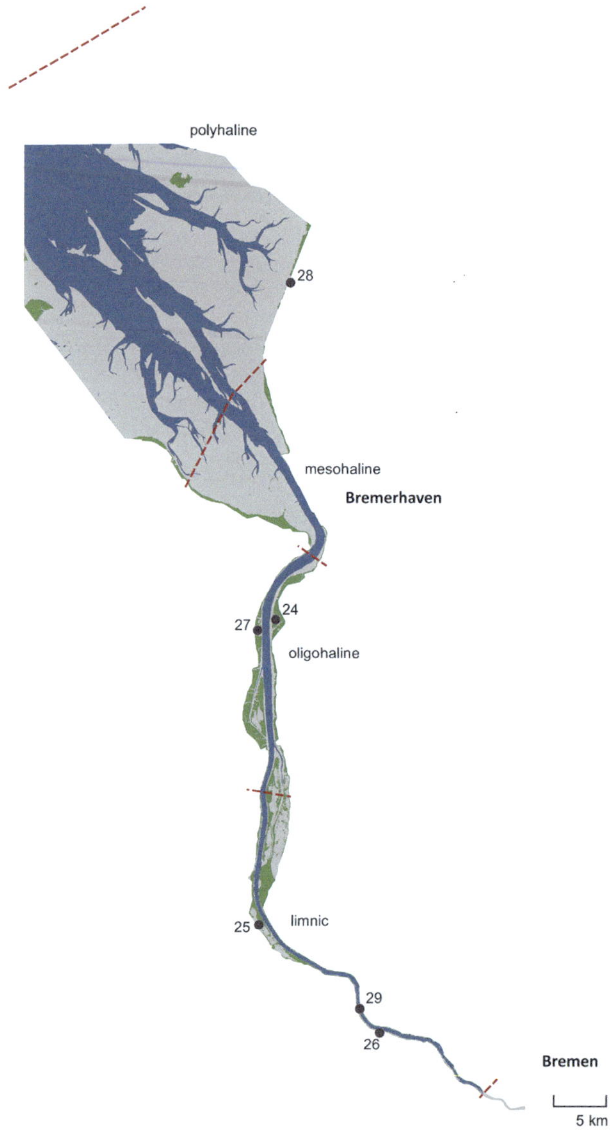

Fig. 5.15 Classification of the Weser according to the Venice system. Figures are indicating measures taken in the estuary in order to improve the ecological (or chemical) status. The enumeration is based on the EU Interreg project TIDE. Source: Geerts et al. (2012)

four parts (see Fig. 5.15). The inland border of the Weser estuary is the weir at Bremen. At the lower Outer Weser Germany's longest container terminal is installed at the harbor of Bremerhaven. All tributaries can be closed-off either by barriers or sluices which have been installed as consequence of flooding during the storm surge in 1962.

Fig. 5.16 Classification of the Schelde according to the Venice system. Figures are indicating measures taken in the estuary in order to improve the ecological (or chemical) status. The enumeration is based on the EU Interreg project TIDE. Source: Geerts et al. (2012)

The Schelde estuary is the longest in the North Sea Region. It extends from the mouth in Vlissingen (NL) to the barrier in Gent (B) about 160 km (Fig. 5.16). The salt water reach is longer than in the Weser and the brackish water zone extends over the harbor of Antwerp. The tributaries of the Schelde in Belgium are not closed by sluices or barriers. The Dutch part of the Schelde is called Westerschelde.

The Elbe estuary has a total length of approximately 140 km from the weir in Geesthacht, south-east of Hamburg, to Cuxhaven at the mouth (Fig. 5.17). At the inland border of the river Elbe lies the port of Hamburg, Germany's largest container harbor. As for the Weser the tributaries of the Elbe have been closed off by barriers and sluices. The brackish water zone is also oscillating depending on the tides and the fresh water discharge from upstream the river – see, for example, Bergemann (1995, 2004).

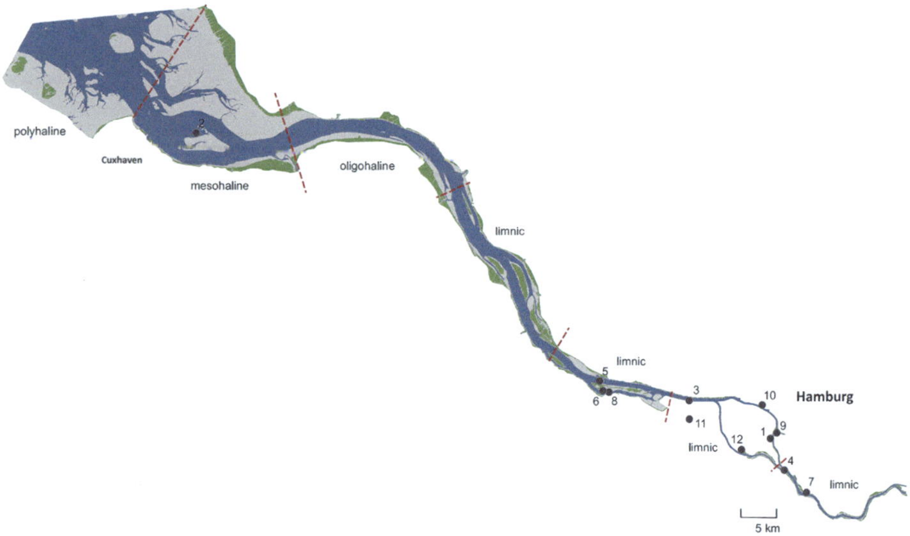

Fig. 5.17 Classification of the Elbe according to the Venice system. Figures are indicating measures taken in the estuary in order to improve the ecological (or chemical) status. The enumeration is based on the EU Interreg project TIDE. Source: Geerts et al. (2012)

Parameter		Elbe	Schelde	Weser
Tidal range	Mouth	2.9 m	3.8 m	3.8 m
	Upstream	3.6 m (Hamburg)	5.5 m (Antwerp)	4.2 m (Bremen)
Fresh water discharge	Mean	722 m^3/s	107 m^3/s	331 m^3/s
	Mean annual min. /max.	247 - 1,709 m^3/s	34 - 253 m^3/s	122 - 798 m^3/s
Current flow velocity	Ebb	0.2 - 0.9 m/s	0.1 - 1.0 m/s	0.1 - 0.6 m/s
	Flood	0.4 - 1.3 m/s	0.25 - 1.5 m/s	0.12 - 1.3 m/s
Length	Barrier to mouth	140 km (Geesthacht - Cuxhaven)	160 km (Gent - Vlissingen)	120 km (Bremen - Bremerhaven)

Fig. 5.18 Characteristics of Weser, Elbe and Schelde. Source: adapted from TIDE Project Consortium (2013)

In Fig 5.18 selected parameters display similarities and differences of the three estuaries. The highest tidal range differs between 3.6 m at the Elbe and 5.5. m at the Schelde. At the Schelde the highest tidal range can be found in Antwerp which is not the inland border. At the Weser the highest tidal range occurs at the inland border, the weir in Bremen, at the Elbe the highest tidal range can be observed at the harbor of Hamburg. The Elbe shows the highest mean fresh water discharge (722 m^3/s) and the Schelde the lowest (107 m^3/s). The current flow velocity either for the ebb or the flood current is very similar.

Fig. 5.19 Different interests and needs in estuaries are linked to different requirements. ©Frank Ahlhorn

In estuaries different interests and needs come together and sometimes lead to conflicts (Fig. 5.19). Some of these interests could mutually be developed, others show contradicting requirements. For example, if the fairway of the estuary shall guarantee the accessibility of an inland harbor it has to be adapted to the current draft of vessels. This affects the shallow water areas and the side arms because of hydrodynamic and morphological changes. Thus, the natural environment of the estuary will be disturbed. Different users have different requirements. This will be exemplified by using the transport sector as relevant driver for changing the shape and structure of estuaries.

5.4.2 Historical Development

Estuaries are used as backbones for trade by ships and vessels. In ancient time sailing boats approached the harbors along the small and narrow estuaries. With the growing requirement for trade the sailing boats were increasing. The consequence was that the journey to inland harbors was more dangerous because of shallow water areas and highly

Fig. 5.20 Ancient map of the Schelde estuary in 1570. In former times the journey from the mouth to the harbor of Antwerp was dangerous due to the islands and mobile tidal flats. Source: Coen (2008)

mobile sand and mud flats. In Fig. 5.20[11] examples of the shape and the structure of ancient estuaries are shown.

The harbor of Hamburg is located 120 km inland of the Elbe mouth and in ancient times the journey took approximately 4 weeks. At the Schelde the situation to approach the harbor of Antwerp was even worser because many islands and sand plates existed. Although the port of Antwerp was important in the Middle Ages, the ability to improve the accessibility was very limited (Fig. 5.20). The advantage of the port of Antwerp in this time was that the goods could be transported 80 km upstream. It was more difficult on land because of missing infrastructure and means of transport. The same developments could be found at the Weser where Bremen, as Hanse City, dominated the sea trade. Due to sedimentation bigger ships were not able to reach Bremen, so in 1827 Bremen bought land from the government in Hannover at the mouth of the Weser and founded Bremerhaven.

[11] Ancient maps have to be taken with a grain of salt because the representation of the real situation could have been influence by various reasons, for example, the emperor would like to see his area of influence bigger than it was or parts of the area have not been visually inspected by the painter.

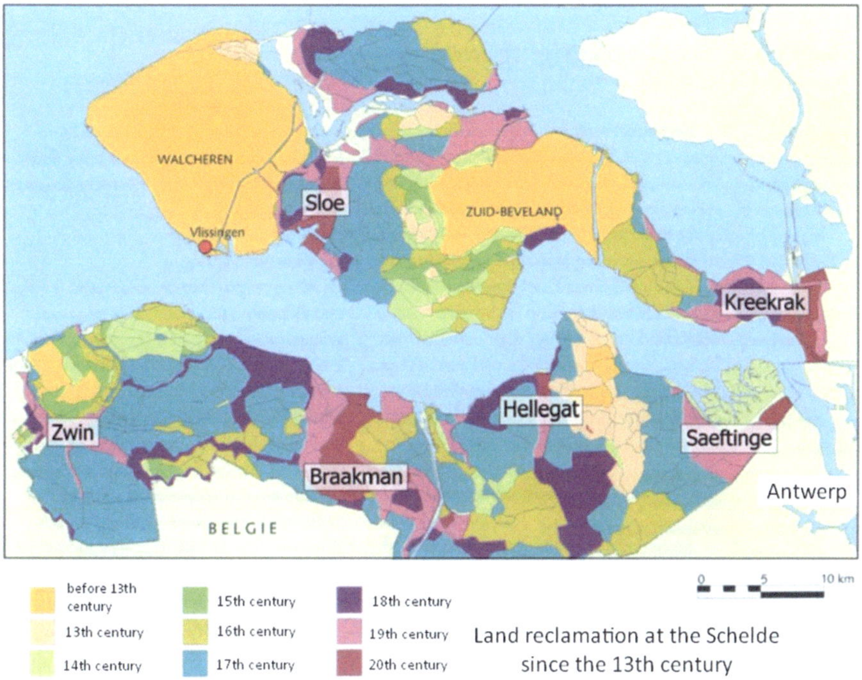

Fig. 5.21 Map of land reclamation works at the Schelde estuary. Most of reclamation was done in the seventeenth century. Also after that period reclamation work was executed due to flooding by severe storm surges. Land reclamation work in the nineteenth and twentieth centuries was due to the strategy of *going forward* and the flood event in 1953. Source: Vroon et al. (1997)

Trade and transport were not the only human activities in estuaries. In Fig. 5.21 land reclamation activities are illustrated. In the Middle Ages land could be reclaimed if and only if the elevation of the area was higher than the mean high water level. As has been shown in Sect. 5.3 by the installation of wind mills and pumping stations low-lying land could be drained under highest efforts. The reclamation of (river) marsh land narrowed the expansion of the water and, thus, reduced the tidally influenced marsh and shallow water areas. The same development can be observed at the Schelde where many storm surges destroyed the small islands and plates in the Middle Ages. Today, the Wester Schelde is an open estuary due to the necessary access to the harbor of Antwerp.

Figure 5.22 illustrates the changes of the cross section of the estuaries from the Middle Ages until today. The wide flood plain has been narrowed by land reclamation and drainage. Dikes protect people and low-lying land against high water levels. The adjacent marshes are intensively used by agriculture or housing and infrastructure for ports. Increasing ship sizes call for deeper fairways.

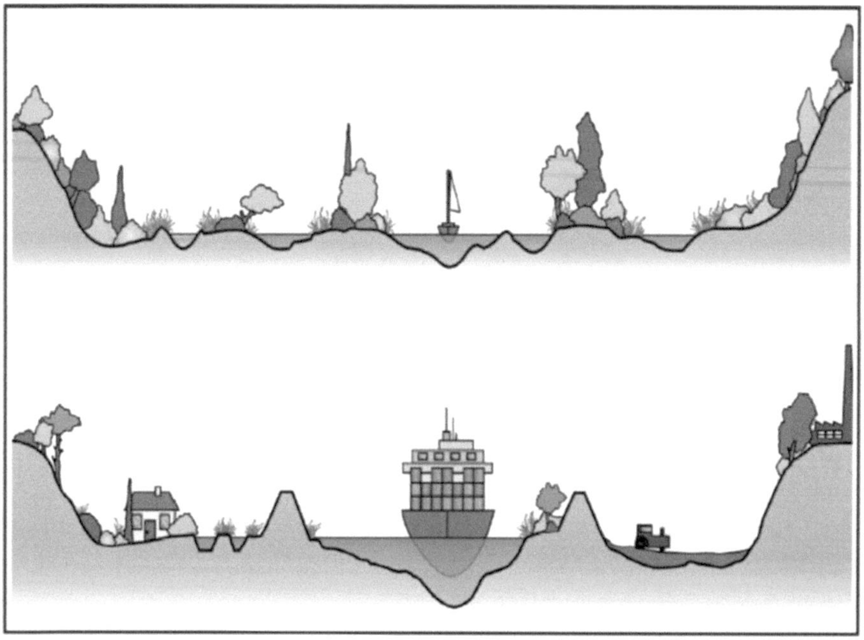

Fig. 5.22 Rivers flow through a wide floodplain and in case of high water levels it was inundated. Side channels and shallow water areas existed (*top*). Change of estuarine shape: deepening, straightening and narrowing of the river floodplain. Consequentially, low-lying land is protected against flooding by dikes (*bottom*). Source: Hamburg Port Authority

5.4.3 Relevant Drivers for Change

The main drivers for change in estuaries are transport and trade, in the past and today. The size of the vessels are continuously growing. In Fig. 5.23 the change of the length is illustrated. From 1950 to 2000 vessels with a length up to 200 m dominated the sea trade. Starting in 2005 the vessels were getting rapidly longer, nowadays, up to 400 m. In parallel, the draft of vessels increased. In former times the draft was at maximum 12 m and nowadays vessels exist with a draft of approximately 16 m. The size of the first vessels was limited by the width of the Panama Canal (PanMax). Today, so-called Post Panmax and Triple E vessels are constructed which are unable to pass the Panama Canal. Therefore, plans have been developed to enlarge the sluices in width and length.[12] Also the Suez Canal restricts the size of vessels in international sea transport not in length but in width (max. 77 m) and height (max. 68 m).

Growing vessel size and international competition demand the adaptation of estuaries to current requirements of transport and, thus, led to continuous human interferences. The fairways in the estuaries have to be continuously deepened.

[12]In summer 2016 the reopening of the Panama Canal with extended sluices should be celebrated.

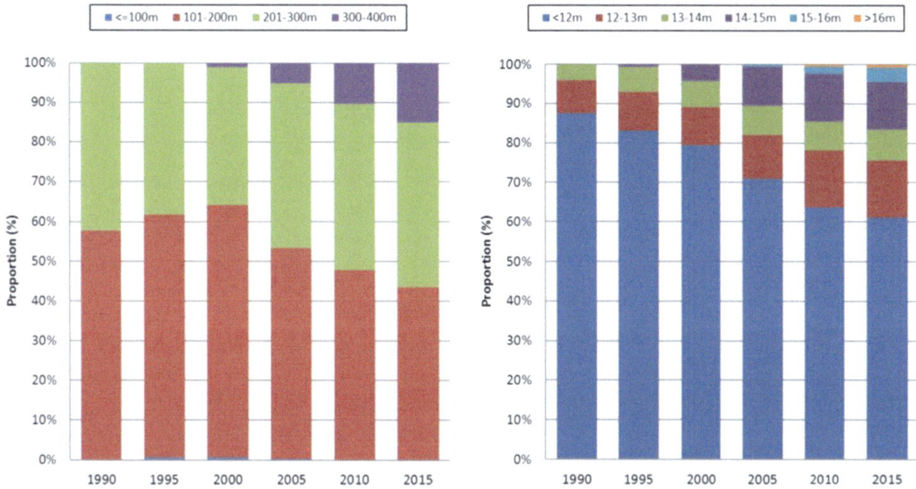

Fig. 5.23 Growing trade demands for bigger vessels. The container fleet increases in draft and length. *Left*: Development of length. *Right*: Changes in draft caused by even bigger vessels. Source: TIDE Project Consortium (2013)

5.4.4 Consequences of Human Interferences

The previously described demand led to the adjustment of estuaries, that is, the main channel for transport (fairway). The adaptation of the fairway to new demands was made by deepening and straightening of the river channels, that is, intervention in the natural morphological development of the river. It is associated with the installation of jetties, groynes and training walls. These impacts should sustain and improve the ability of the current to minimize maintenance work of the fairway. Generally, the rivers along the southern North Sea Region are ebb dominated, that means that the ebb current is stronger and its duration is slightly longer in contrast to the flood current. Thus, sediment is mainly transported downstream the river/estuary toward the open sea. The narrowing of the main channel leads to a higher current and, thus, to a higher energy for sediment movement. Straightening and deepening of the fairway has to be accompanied by hard constructions to fix the planned channel at its position and to minimize the impact of higher currents pose on the embankment. These impacts can especially be observed in low-lying marsh areas such as in the southern North Sea Region.[13]

[13]For example, Blackbourn (2006) described the ancient river engineering work at the Rhine where a similar development could be observed.

Fig. 5.24 List of fairway deepening at the river Weser starting in 1887. Source: Lucker et al. (1995) and www.wsa-bremerhaven.wsv.de

Year	Draft [m]
1887 - 1895	5.00
1913 - 1916	7.00
1925 - 1928	8.00
1953 - 1959	8.70
1973 - 1977	11.00
Currently planned	12.80 (Bremerhaven to Brake)
	11.10 (Brake to Bremen)

Morphological Development

The morphological development mainly caused by human interferences is exemplified by the Weser and the Schelde.

In Fig. 5.24 a list of deepening and fairway adaptations at the Weser over the last 130 years is shown. The natural variations of water depth and the oscillation of the navigation channel led to the idea of the first Weser correction at the end of the nineteenth century – see Franzius (1888/1991). The principal objective was to straighten, to deepen and to fix the fairway of the Weser at one channel. This should support the self-sustaining hydrodynamic and morphological changes proposed by Franzius to enable ships with an increased draft to approach Bremen. One channel was chosen being the fairway and, therefore, jetties, groynes and training walls were constructed. The following steps of deepening and straightening were to some extent consequences of the first correction and to some extent owed to the continuous increase of ship sizes (see Fig. 5.24).

To illustrate the difference before and many years after the correction of the Weser estuary, compare the situation of Figs. 5.25 and 5.26. The harbor of Bremerhaven is located at the bottom right corner on the east side of the fairway. According to the theory of Homeier (1969) the two main channels at the mouth of the Weser were oscillating, that is, while the northern channel was broad and deep the southern channel was narrow and shallow and vice versa. The oscillation showed an estimated frequency of 400–500 years (Homeier, 1969). Caused by the solidification and the deepening the fairway was directed to the east embankment. In the course of time due to lower current and eventually higher sedimentation rates the sand flats were growing. This was mainly caused by these severe interventions. The natural dynamic of the mouth system was further disturbed by land reclamation and by building dams into the Wadden Sea area to avert harmful currents from the southern embankments.

The situation in 1650 showed a meandering channel system intersected by sand and mud plates. Various side channels existed with shallow water areas. The shape and

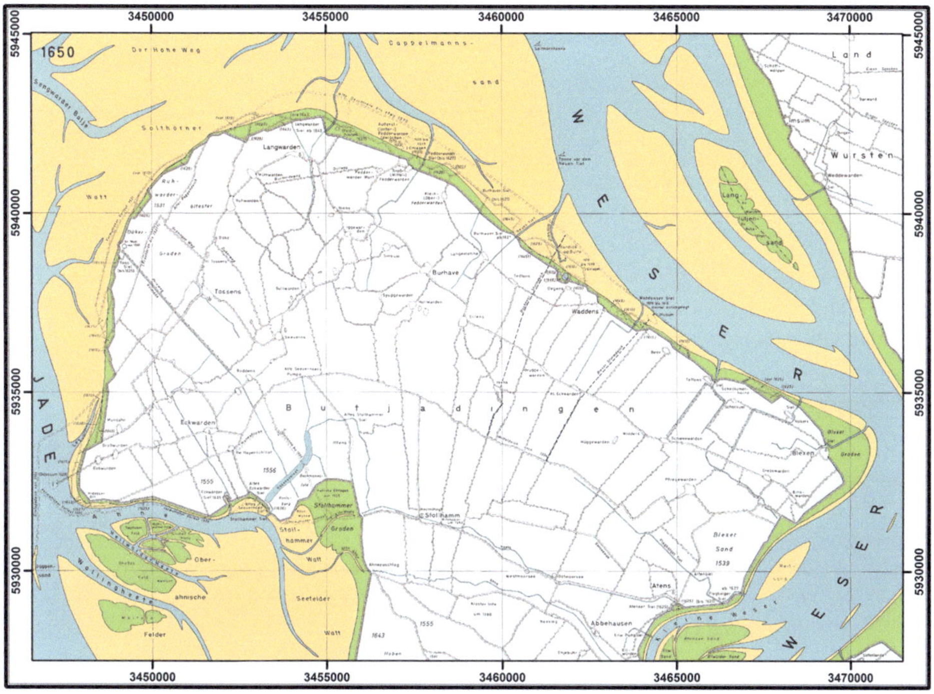

Fig. 5.25 The Weser at Bremerhaven showing a multichannel system with intersection of sand plates and small islands in 1650. Source: Homeier et al. (2010). (Published with kind permission of ©Niedersächsischer Landesbetrieb für Wasserwirtschaft, Küsten- und Naturschutz 2010. All Rights Reserved).

structure was dramatically altered until 1960 (Fig. 5.26). Most side channels were filled up or reclaimed. The main channel of the Outer Weser was straightened and deepened.

The same development can be observed further upstream. Figure 5.27 illustrates the results of different interventions in the river channel. Compare the situation shown in Fig. 5.27a, b. The natural dynamic channel system was changed to a channelized fairway. Only two islands still remain in the Lower Weser Harrier Sand and Strohauser Plate. In total approximately 120 km of embankments were removed through these interventions (Lucker et al., 1995).

In Fig. 5.28 the changes in bottom depth of the river Weser illustrated these interventions. In 1887 the bottom depth at the weir (left side) was nearly the same height as the sea level. After the correction and the dredging works the bottom depth decreased to approximately –4.50 m below sea level. At the same time the river bed depth at the mouth in Bremerhaven (right side, 70 km) was deepened to –14.5 m.

In parallel the cross section of the fairway was broadened. Figure 5.29 displays the changes of the cross section over time. It shows the time line of change from 1887 until

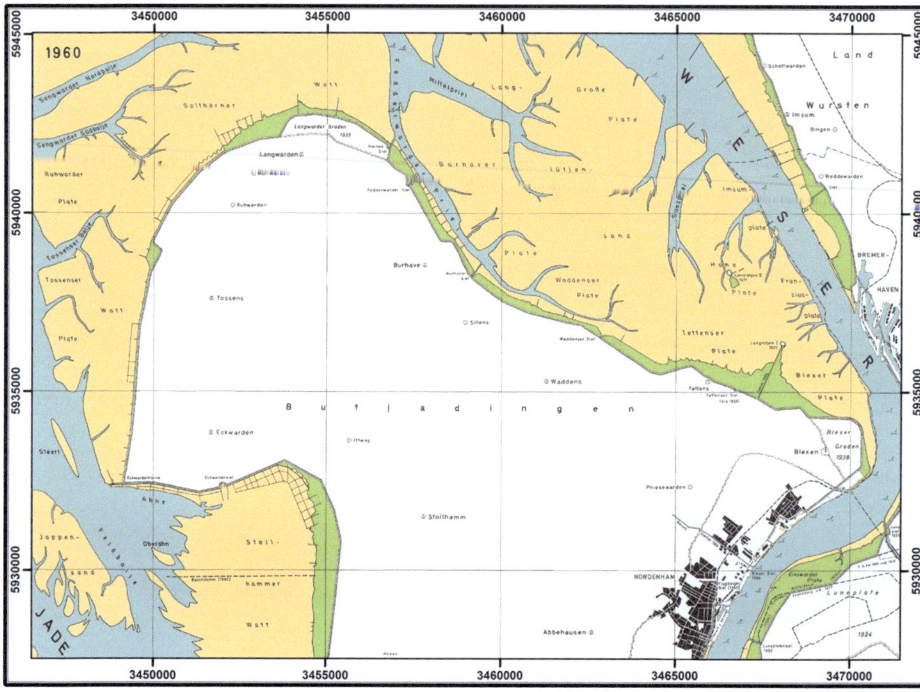

Fig. 5.26 The Weser at Bremerhaven after several corrections and adaptation of the fairway until 1960. The sand plates and mud flats have increased in front of the peninsula of Butjadingen and the fairway is fixed at the northern side of the estuary in front of the quays of the city of Bremerhaven. Source: Homeier et al. (2010). (Published with kind permission of ©Niedersächsischer Landesbetrieb für Wasserwirtschaft, Küsten- und Naturschutz 2010. All Rights Reserved).

1978. The depth was increased from near sea level to –11 m and the width of the fairway in 1978 was more than twice as of 1887.

A similar development could be observed at the Schelde. During the last 100 years interventions such as reclaiming land, dredging and the installation of river conservancy means changed the shape of the Schelde. Figure 5.30 displays the change from 1959 to 2012. The multichannel system still exists, but changed at some stretches to a double channel system. The fairway was broadened (Fig. 5.30 bottom) and deepened. Some of the mega sand flats (mega ripples) are still existing.

A visual comparison between Weser and Schelde shows that the Weser is more channelized than the Schelde. The former meandering channel system of the Weser does not exist anymore, it was reduced to a single channel. At the Schelde the multi-channel system was, at certain stretches, reduced to a double channel system. The double channel system is still existing because the ebb- and flood current flows through different channels in the Schelde.

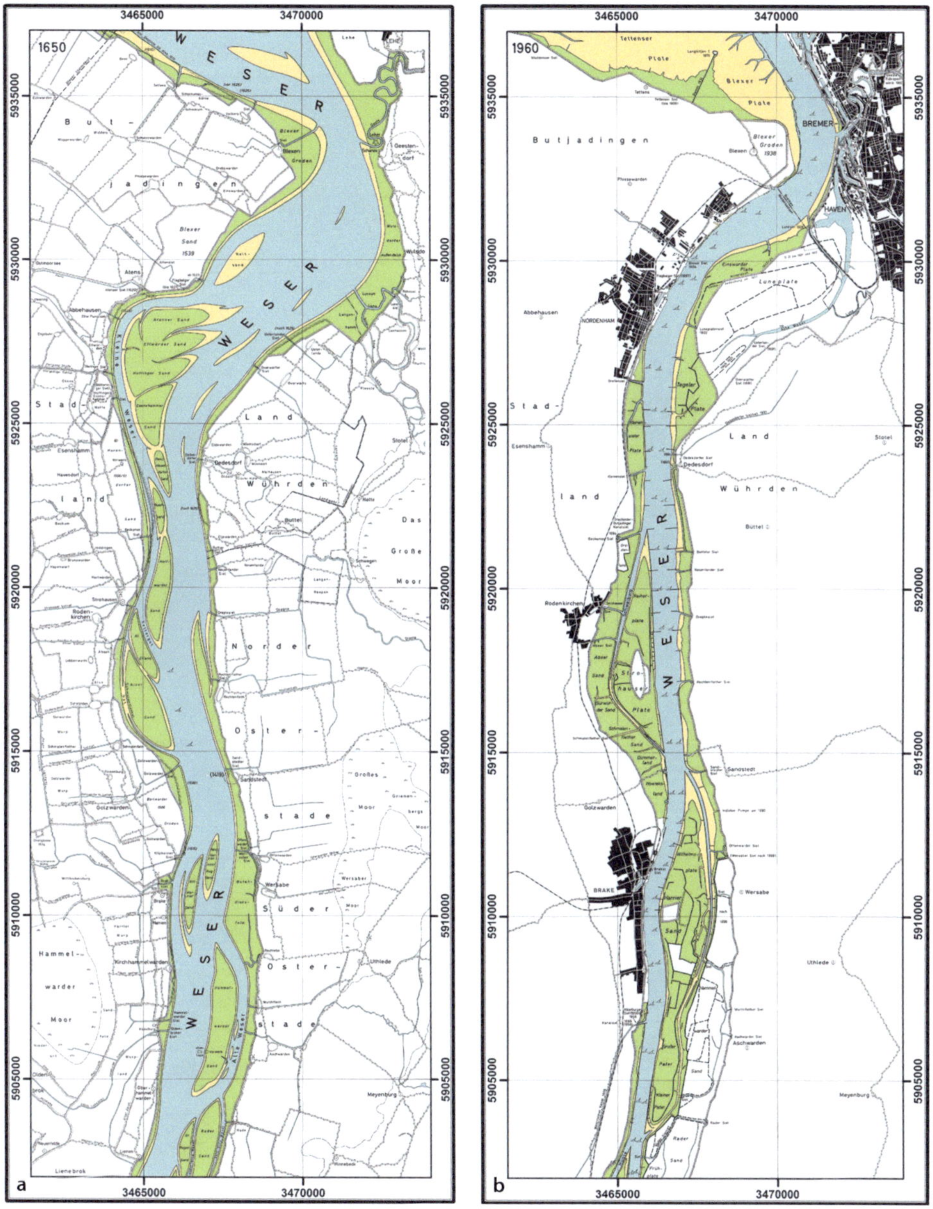

Fig. 5.27 Lower Weser between Bremerhaven and the city of Brake. Source: Homeier et al. (2010). (Published with kind permission of ©Niedersächsischer Landesbetrieb für Wasserwirtschaft, Küsten- und Naturschutz 2010. All Rights Reserved).

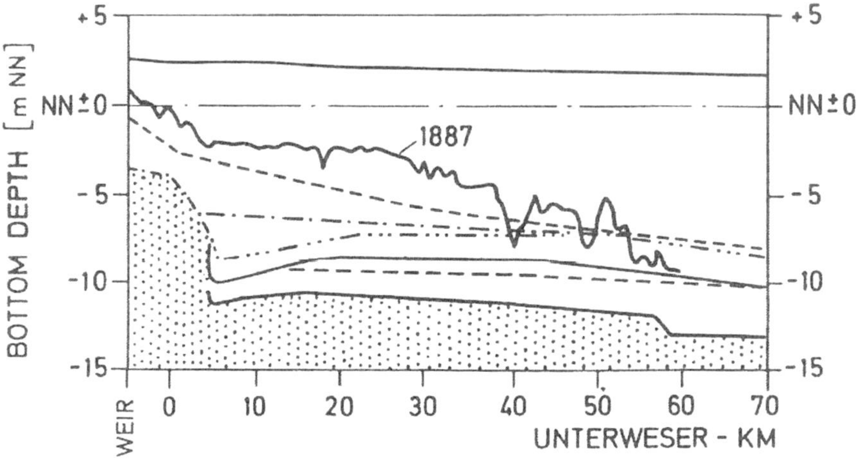

Fig. 5.28 Bottom depth of the Weser estuary from the weir in Bremen to the mouth in Bremerhaven. The *black line* indicates the bottom depth in 1887, the *dashed lines* show the changes in bottom depth by dredging. The lines in between mirror the steps of deepening shown in Fig. 5.24. Source: Kunz (1995)

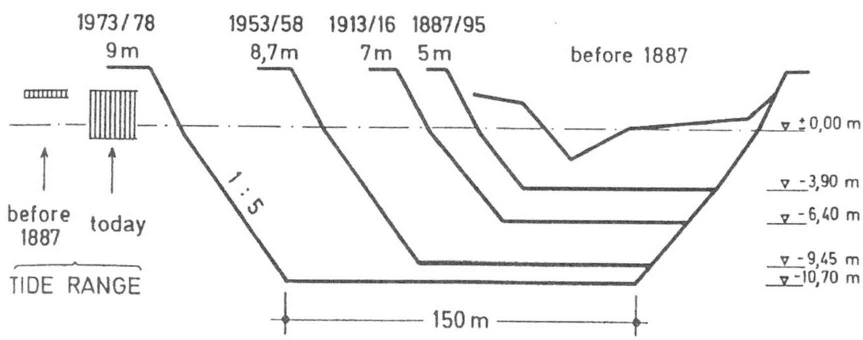

Fig. 5.29 The enlargement of the river bed width from 1887 until 1978. Source: Wetzel (1988) adapted by Kunz (1995)

Water Level Changes

One consequence of the river conservancy works of the Weser was the change in water levels. Figure 5.31 illustrates the change upstream at the weir in Bremen (black line). While the mean high water level decreased slightly in approx. 100 years, the low water level decreased from near 3 m above sea level to approximately –1.50 m below sea level in 1980. Subsequently, at the weir the tidal range increased from about 0.2 m up to 4.2 m today. In comparison tidal range at the mouth in Bremerhaven did not significantly change (dashed line, Fig. 5.31).

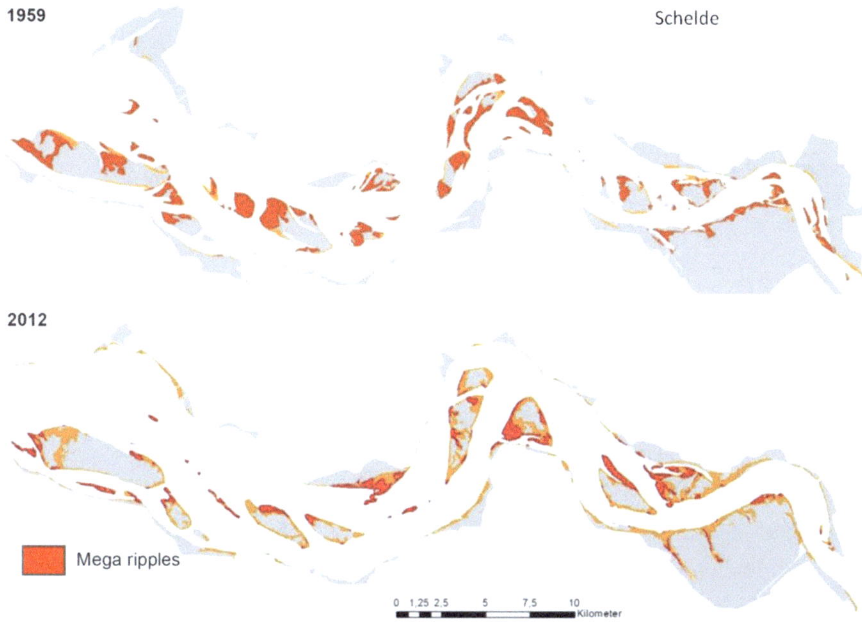

Fig. 5.30 The morphological changes of the Schelde channel system and the illustration of the behavior of the mega ripples within the fairway. Source: IAHR (2015)

Fig. 5.31 Changes in water level (*MHW*: mean high water, *MLW*: mean low water) at the weir (Bremen, *black line*) and at the mouth (Bremerhaven, *dashed line*). Insignificant changes of water levels at the mouth, but significant changes of the MLW at the weir in Bremen can be observed. Source: Kunz (1995)

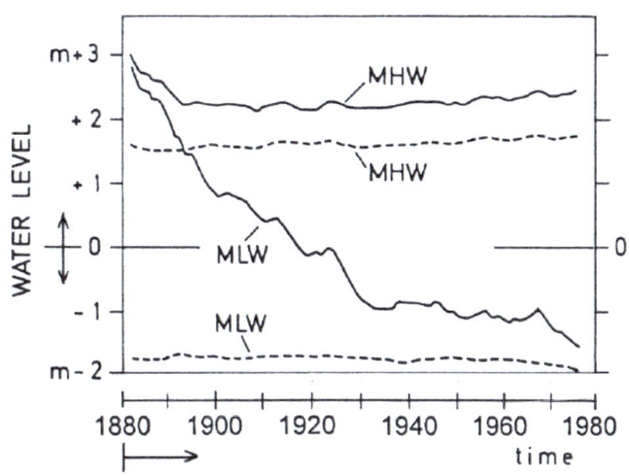

The development and distinct change in water levels is not solely observable at the Weser. Figure 5.32 displays the change in tidal range at the Elbe and at the Schelde rivers. The reasons for these changes are similar. The tidal range in Hamburg has almost been doubled in the last 100 years (1.8 m in 1870, Fig. 5.32) and at the Schelde in Antwerp tidal range has been increased by somewhat more than 1 m.

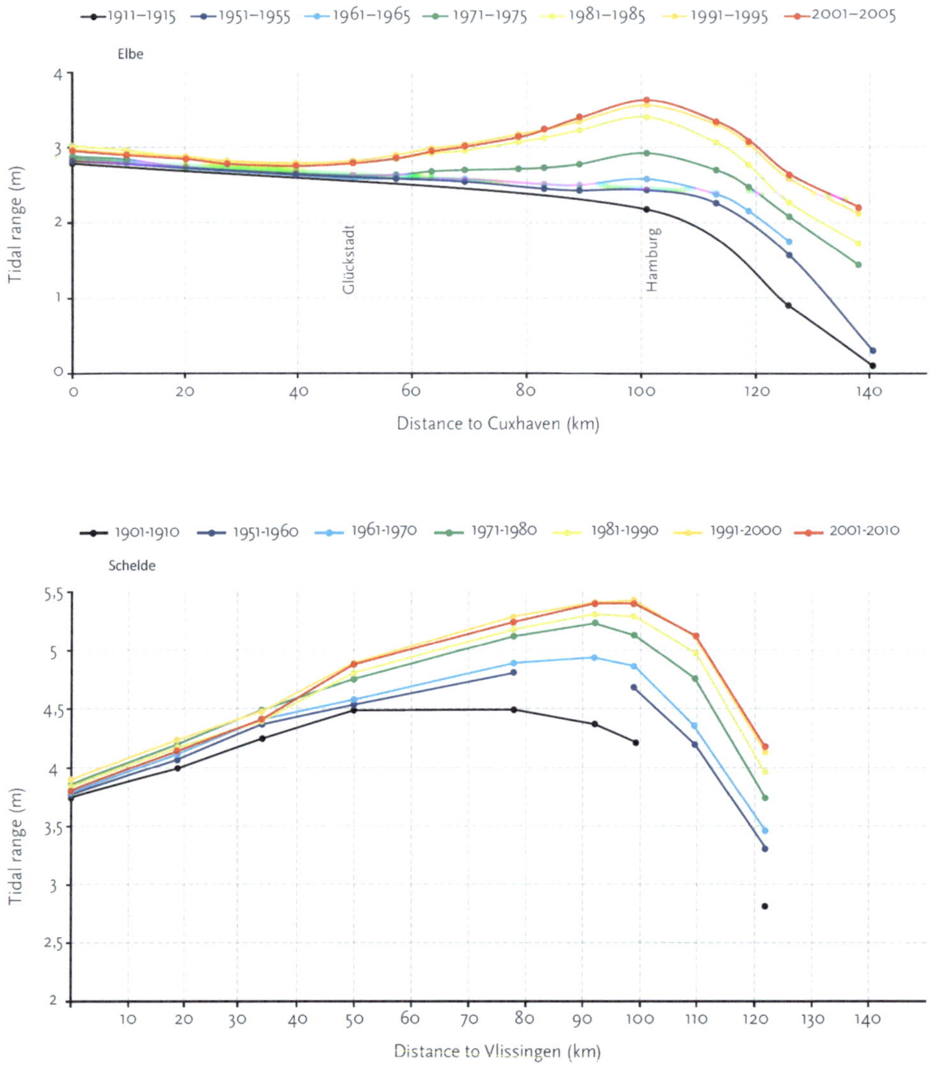

Fig. 5.32 Changes of water levels in the Elbe and Schelde estuary. At the Schelde the tidal range increased from 4.1 m to 5.5 m in 90 years at Antwerp (*bottom*, km 90). At the Elbe tidal range increased in Hamburg from 2.1 m to 3.6 m, at the barrier in Geesthacht tidal range increased about 2 m in 90 years (*top*, km 140). Source: TIDE Project Consortium (2013)

The interventions led to increased current velocity and to an increase of the tidal range at the inland border in the Weser river. As consequence the brackish water body was shifted more upstream. While in former times the brackish water zone oscillated a few kilometers depending on tides and fresh water discharge, today the oscillation extends more than 10 km. It is important to keep in mind that these estuaries were not solely affected by human interventions but also by climate change, that is, sea level rise. The effect of sea level rise also has an impact on water levels, the oscillation of the brackish water zone and so on. Therefore, it is crucial to understand the natural dynamic and variability and the influences of human interventions on these processes. In most cases it is difficult to differentiate if these impacts are based on human interferences or natural variation or if they are a mixture of both.

Nevertheless, tidal range is a proxy to show that changes have taken place in estuaries, further parameters are also important such as current velocity, time of the tide capsizing point and many more.

Dredging Activities

Anthropogenic morphological changes of river channels are generated by dredging and other river conservancy works such as training walls along channels. The question is what to do with the sediment? Two main strategies exist in sediment handling. One strategy is to relocate the sediment from certain reaches or channel parts to another part where the depth is insufficient for shipping and transport. This material is, for example, relocated on spots where undesirable erosion occurs. The second strategy is the removal of sediment from the river, mainly sand and gravel which can be used e.g. for infrastructure projects. Prior to the use of this material on land it has to be tested for contamination with heavy metals and other pollutants.

This section exemplifies the different strategies and the amount of dredging in two estuaries. At the rivers Weser and Elbe the removal of sediment prevail either to be used by third parties for infrastructure projects or by disposing the sediment on dumping sites in the North Sea. At the Schelde the sediment is relocated to certain spots in side channels (see Sect. 6.2.3).

Figure 5.33 displays the amount of dredged sediment in five distinct sections of the Elbe river. At the mouth the amount of dredged material increased between 1999 and 2009. Unfavorable morphological changes (sediment accretion) demanded for higher efforts. The amount of dredging in the port of Hamburg showed reasonable amount as well. Mainly fine grained sediment has to be dredged. A superficial explanation for increasing dredging activities is that many estuaries have changed their tidal characteristics, that is, tidal pumping is prevailing. For the Elbe river net sediment export has changed by different reasons but one of them is related to the changes due to river engineering works and reduced fresh water discharge from upstream.

Figure 5.34 illustrates the increase of dredging between 2000 and 2009: from 15 Mio. m^3 in 2000 up to more than 25 Mio. m^3 in 2008. Efficacy of dredging plays an important role. There is a distinction between maintenance and capital dredging. Capital dredging is applied to achieve a certain depth of a fairway, maintenance dredging is necessary to keep the fairway or channel on a desired depth and shape.

In comparison at the Weser the amount of dredging is less (Fig. 5.35, mean 10 Mio. m^3). In 2006 the amount of dredged material was higher because a turning point for big vessels was installed (total length of approximately 2 km and width of 600 m), approximately 2 Mio. m^3 of sediment were relocated (Krämer et al., 2010). The reason

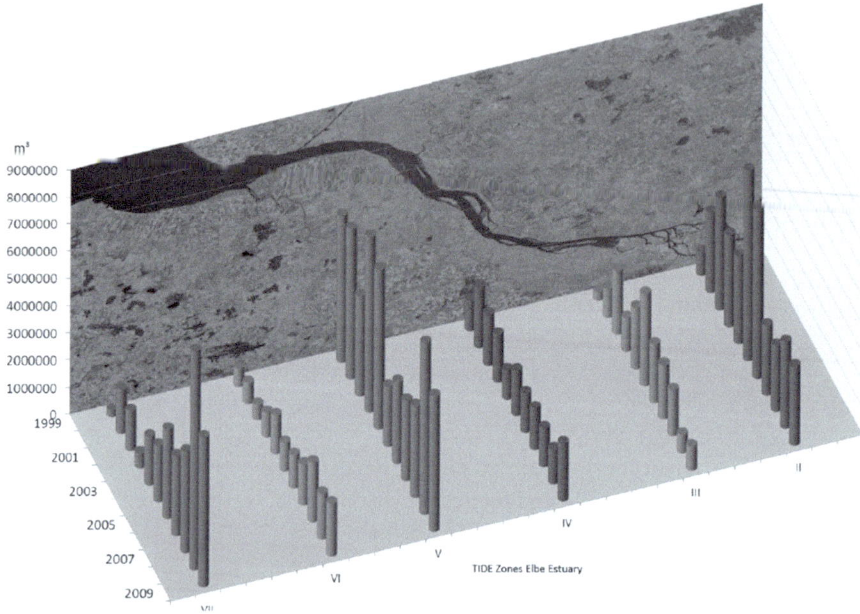

Fig. 5.33 Example for dredging amounts according to different reaches of the Elbe estuary (1999–2009). Source: HPA and WSV (2012)

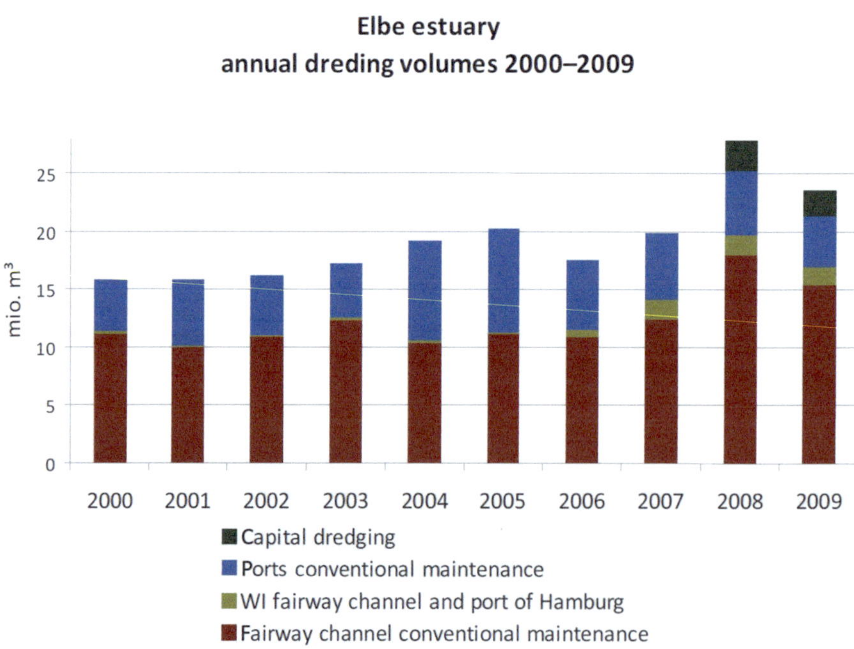

Fig. 5.34 Total amounts of dredging at the Elbe estuary from 2000 to 2009. Source: BioConsult (2013)

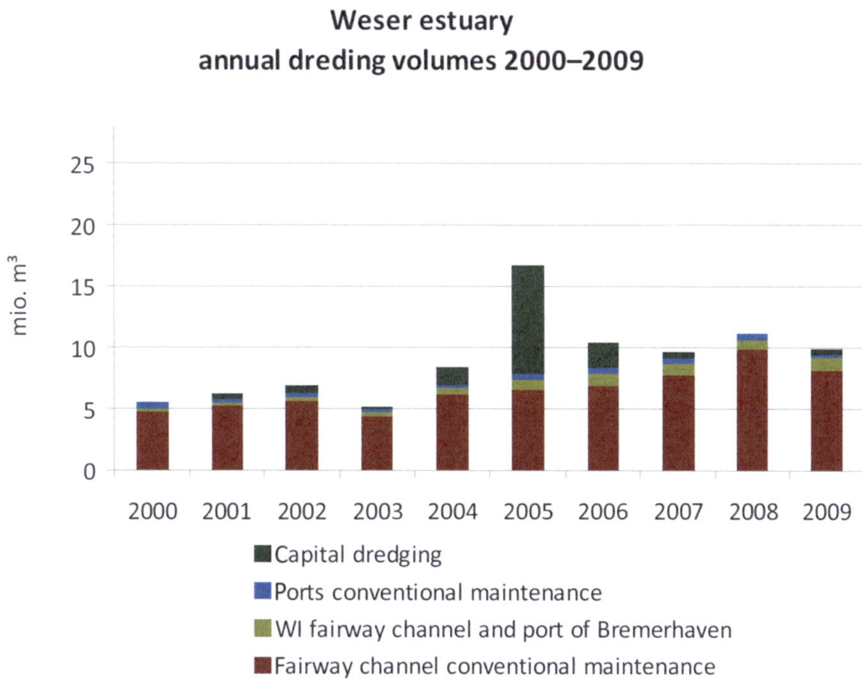

Fig. 5.35 Total amounts of dredging at the Weser estuary from 2000 to 2009. Source: NLWKN and BioConsult (2012)

for the high amount of dredged material in 2005 was the extraction of sediment by third parties (NLWKN and BioConsult, 2012).

In contrast to the Elbe at the Weser the highest efforts for dredging are located in the mouth (Fig. 5.36). The illustration shows that at the Weser in upper reaches less dredging is necessary (2000–2006). One reason is that the Weser does not show the effect of tidal pumping up to the inland border. The lower parts of the Lower Weser have to be dredged because the current has to pass the so-called Blexer Bogen where the current follows a curve. As for natural meandering channels a slip-off slope and cut bank exist. At the slip-off slope sedimentation rates are high which affects the harbor infrastructure at this side of the embankment. In the upper reach the current is strong enough to flush the sediment downstream. Therefore, dredging activities at the harbor of Bremen are lesser than at the harbor of Bremerhaven (Fig. 5.36). On the other hand, the harbor of Bremerhaven is more important for trade than upstream harbors which are smaller in size and goods turnover.

Despite this fact the side channels of the two remaining islands (Strohauser Plate and Harrier Sand) are filling up due to higher current velocity in the main channel.[14] Although

[14]The water is penetrating almost mutually from the north and the south into these channels. The converging currents reduce in speed and sedimentation takes place.

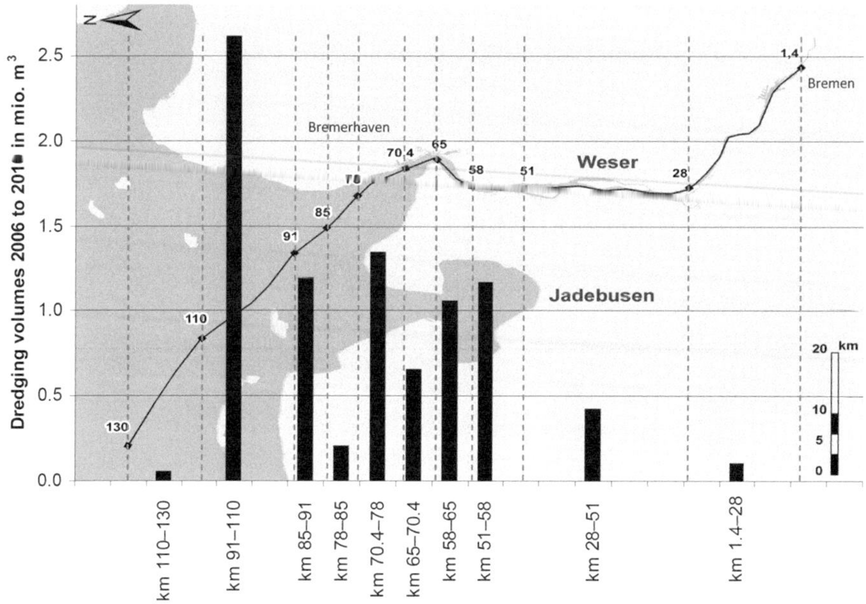

Fig. 5.36 Amounts of dredging along the Weser estuary from 2000 to 2006. In the background the map shows the extension of the Weser between the harbor of Bremerhaven at the mouth and Bremen. The figures at the black line indicate the river-km from the weir in Bremen up to the border of the Outer Weser. Source: NLWKN and BioConsult (2012)

a decreasing low water level is beneficial for drainage of adjacent marsh lands, it shows obverse effects by filling up side channels. As long as drainage facilities are directly connected to the main channel this effect is beneficial, but when these facilities drain into side channels drainage is getting worse.

5.4.5 Identification of Challenges and Problems

In the previous sections the changes and developments until today have been outlined. These changes were exemplified through the adaptation of the estuaries to new demands for transport and trade. This results in the continuous deepening and further intensification of river conservancy works. In recent years other interests and needs were getting equally important because interest groups or changes in legislation urged EU Member States governments to consider them, for example, NATURA 2000 or the Water Framework Directive (WFD). For example, through the EIA plans and programs have to be assessed against impacts to the environment. Therefore, also plans for further deepening or adaptation of fairways have to run through these procedures. In official documents such as the Integrated River Basin Management Plans[15] according to the WFD deficits and measures to improve existing and identified problems are listed.

Here, reference is made to two recently conducted research projects for European estuaries. On the one hand, the Tidal River Development project (TIDE[16]) running from 2010 to 2012 in the framework of the EU Interreg IVB program dealing with the identification and assessment of measures to improve the status (physical and environmental) of European estuaries. On the other hand, the Estuaries on the move project (EMOVE[17]) also conducted in the framework of the EU Interreg IVB program. The aim of EMOVE was to identify (integrated) solution approaches and to find a way on how to bring/motivate stakeholders in respective estuaries to become shareholders. Shareholders in the way to jointly work on challenges, problems and development goals for parts or the entire estuary.

For the Weser river the identification process was conducted by an expert group judgment. Although, the expert groups in the EMOVE and TIDE projects did not have any overlap, the general result is similar: navigation or transport/logistic shows negative impact on the natural system. Within the EMOVE project it was based on the application of the DPSIR approach. DPSIR is the abbreviation of drivers (or driving forces), pressures, state, impacts and response and is a causality based concept to highlight the interrelationship between human activities and environmental degradation (Pirrone et al., 2005). The definition of the terms is as follows:

- The *driving forces* are processes and anthropogenic activities (production, consumption, recreation, etc.) able to cause pressures
- The *pressures* are the direct stresses, deriving from the anthropogenic system, and affecting the natural environment, that is, pollutant release
- The *state* reflects the environmental conditions of natural systems (air, soil and water quality)
- The *impact* is the measure of the effects due to changes in the state of environmental systems
- The *response* is the evaluation of actions; oriented to solve environmental problems in terms of management strategies

This approach was and is wide-spread applied – see, for example, in OECD (1997); Smeets and Weterings (1999). Based on expert group judgment per estuary relevant pressures according to a list of drivers were identified. The result for the Weser is shown in Fig. 5.37.

In order to identify relevant pressures of an estuary which are the direct stresses posed by human interference several driving forces have been suggested. A long list of drivers

[15] See Sect. 6.1.1.

[16] www.tidetoolbox.eu

[17] www.emove-project.eu

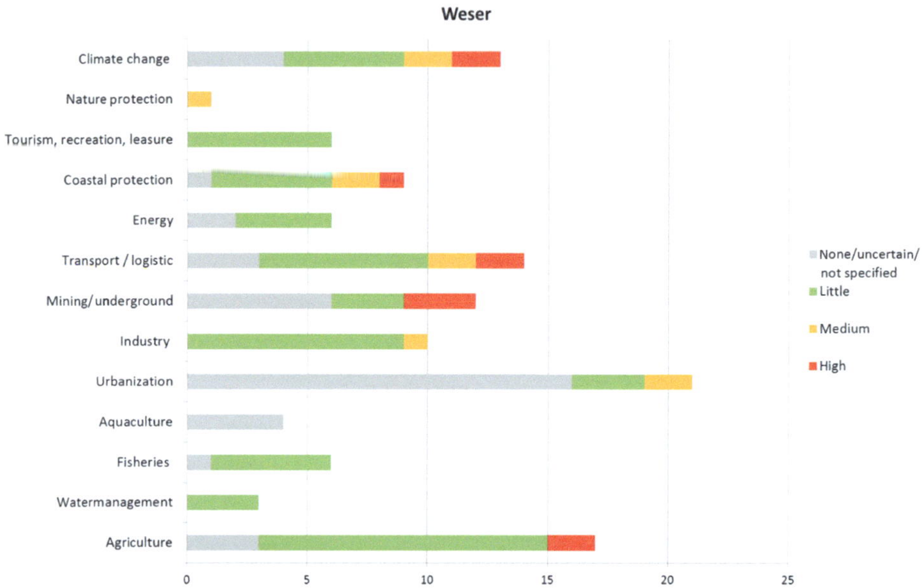

Fig. 5.37 Identification of relevant Drivers and Pressures at the Weser. Source: EMOVE Project Consortium (2015)

has been assessed and reduced to a short list of relevant drivers applicable for the respective estuary. For example, the drivers applicable for the Weser are shown in Fig. 5.37. The task of the expert group was to assign pressures to these drivers. For example, the driver transport/logistic is linked to twelve pressures such as infrastructure and changes in morphology. That means the transport sector needs infrastructure such as harbors or sluices to be able to process container vessels. This infrastructure needs the natural resource *space/soil* (here: marsh land) to be built. The status of the natural area has to be changed, namely destroyed, that is, paved by asphalt, etc. Regarding the pressure changes in morphology the relations have already been touched in the previous section.

Furthermore, Fig. 5.37 shows the relevance of these pressures, that is, how strong the impact of the respective driver is estimated. The transport/logistic driver shows according to the expert judgment, high relevance to three pressures such as commercial harbors, salt water intrusion and dredging activities. These pressures assumed to be of highest relevance if the effect of natural degradation is considered, the pressures as consequence of the respective driver have the most impact on the natural environment. Coastal protection and climate change show also high relevance in pressures which have severe impact to the natural environment (EMOVE Project Consortium, 2015).

The identification of pressures indicates that something is unbalanced in the estuary. Subsequently, the pressures can lead to or are triggers for various kinds of conflicts. Within the EU Interreg project TIDE the conflicts were investigated in detail – see Cutts and Hemingway (2013). The matrix in Fig. 5.38 provides an overview on the

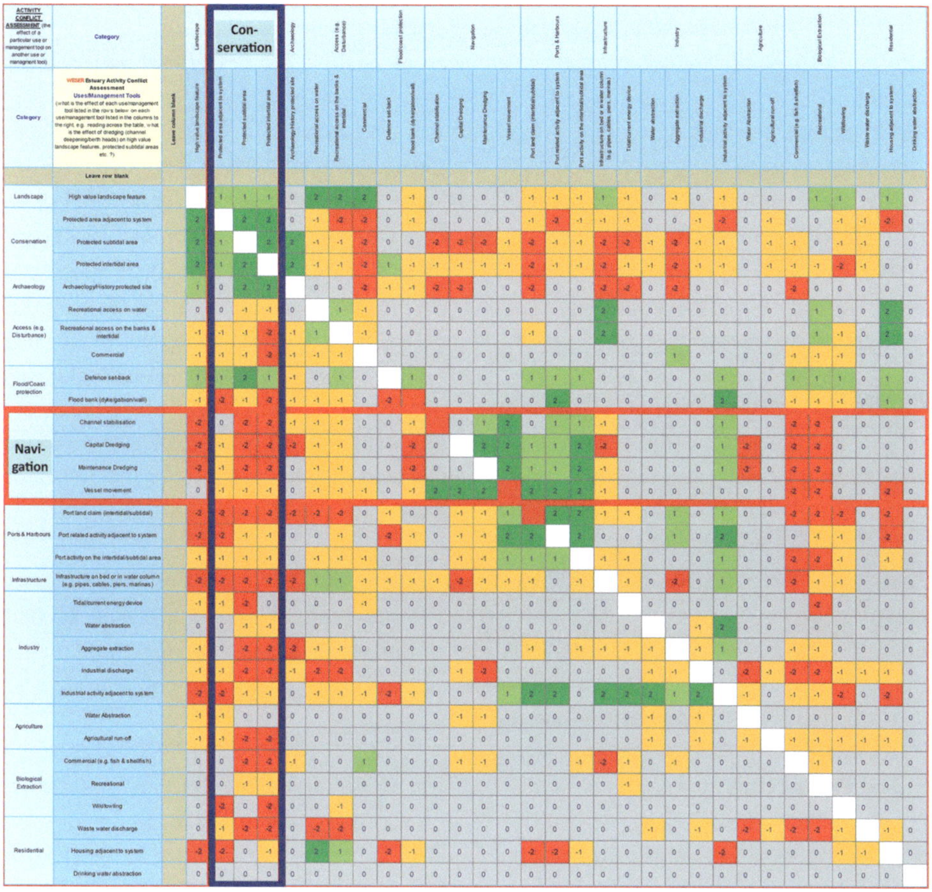

Fig. 5.38 Conflict matrix for the Weser estuary as identified within the TIDE project. Source: adapted from Cutts and Hemingway (2013)

indication on conflicts according to several uses and management tools at the Weser estuary.

Within in the TIDE project different uses or management tools were identified and assigned to a specific category. For example, the category navigation contains the uses/management tools *channel stabilization*, *capital dredging*, *maintenance dredging and vessel movement* (see Fig. 5.38 red frame). The conservation category contains *protected areas adjacent to system*, *protected sub tidal areas* and *protected inter tidal areas* (Fig. 5.38 blue frame). Now, the negative scores (represented by red boxes) indicate that the vertically listed uses/management tool has negative effect on the horizontal uses/management tool. *Channel stabilization* has an impact on the *protected sub tidal and inter tidal areas* (Fig. 5.38 blue and red frames). Also the *dredging activities* have been assessed to have negative effects on the *protected sub tidal and inter tidal areas*. On the one hand, *vessel movement* is seen as not so worse for the *protected areas* (yellow boxes). On the other hand, *dredging activities* are seen as positive for *vessel movement* (green boxes).

In general, the matrix could be interpreted that the red boxes indicate severe conflicts, yellow boxes negative impacts but less severe. Green and light green boxes indicate positive influences of uses and/or management tools. Grey boxes indicating no or neutral impact on each other. The conflict matrix shows that most conflicts exist between navigation and other uses (red frame) and that nature conservation is mostly threatened by all other uses (blue frame). This statement is neither remarkable nor new but visually indicates that the main driver for changes in estuaries is linked to transport. The situation is slightly different for the other estuaries under investigation in the TIDE project – see Cutts and Hemingway (2013). The important step is now to consider this result when planning new adaptations of fairways posed by an increasing demand of the transport sector. The integration of the consequences either on the natural environment (degradation) or for consequential costs to reduce, to minimize or to avoid negative effects.

Example: Current Situation of Water Management at the Weser With regard to the previous section the current situation for water management in general will now be outlined. If the transport sector is one of the major drivers one key for solving conflicts lies in the comprehensive and integrated treatment of future plans and a thorough reprocessing and reassessment of past activities and their consequences.

Flood Protection

As an example, in Fig. 5.39 the flood protection measures at the Weser after the severe storm surge in 1962 are illustrated. All tributaries were closed off either by barriers or sluices. The concept was to protect the low-lying marsh lands against flooding by storm tides. In case of a storm surge these barriers and sluices will be closed and the high water levels won't penetrate further inland. But, storm surge events sometimes are accompanied by heavy rain fall events which led to a higher run-off in the hinterland. Caused by the closed barriers the inland water can not be drained if necessary, thus, the water level in the discharge system is increasing. The risk of flooding by fluvial high water could increase. Therefore, additional flood control measures have been installed behind the closed barrier, for example, retentions ponds that could be flooded during high water to protect covered area against flooding. The same development can be observed at the rivers Schelde and Elbe. Side channels either were closed off or reduced in their shape, barriers and sluices are controlling the water level of the tributaries. Sea walls and dikes have been built and are continuously heightened along the estuaries.

Drainage and Watering of Low-Lying Areas

The consequence of flood control measures is the necessity to install infrastructure to drain the low-lying marsh land. At some places these features are solely tidal gates, sometimes

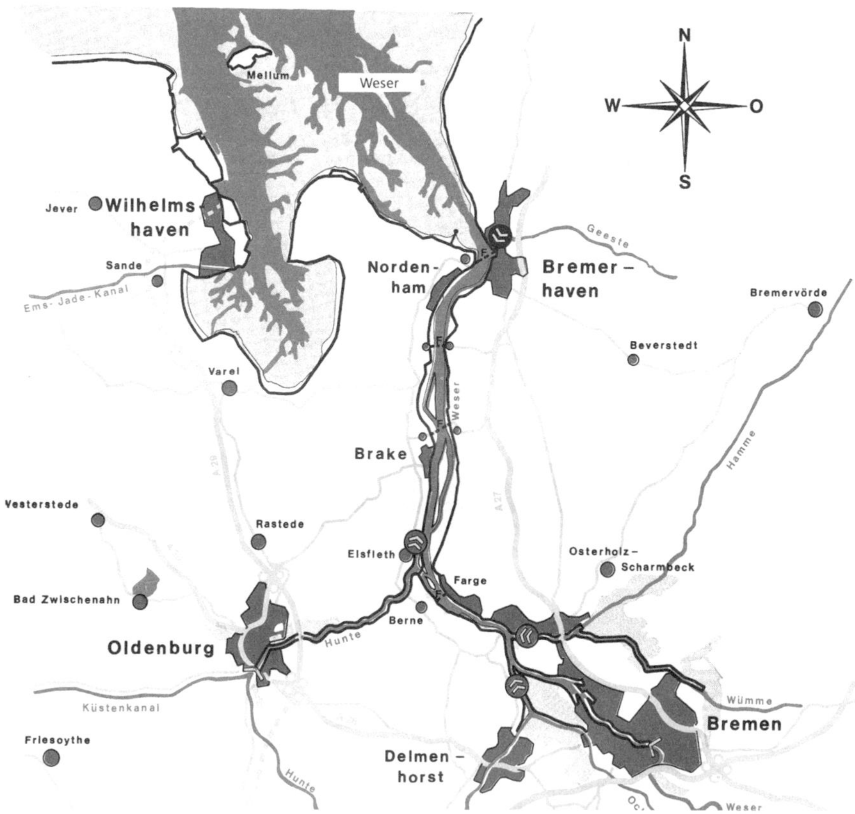

Fig. 5.39 Flood protection at the Weser after 1962. The solid black line along the river indicates the improvement of river dike.The tributaries were closed by tidal barriers to control upstream water levels and prevent floodings. Source: MFAF and SIB (1979).

they are combined buildings together with (mechanical) pumping stations. In the northern part of the Weser estuary a special situation exists. The drainage management features are also able to bring fresh water into the area (watering). These features have been installed as consequence of the first Weser correction because an upstream movement of the brackish water zone was anticipated. Since decades, farmers flush their ditches with fresh from the Weser to feed their cattle. Today, the water quality is temporarily not sufficient, thus, the cattle have to be supplied with drinking water. The drainage system in this low-lying coastal area is very sensitive to higher water levels in the river Weser. If the high water in the river avoids the instantaneous drainage of inland water, at present no retention ponds exist (more general: precaution strategy) to temporarily store the water. The storage capacity of the ditch and canal system was calculated decades ago and is limited.

Ground Water Body

The situation that fresh water from the river Weser is needed is also caused by the saline ground water body underneath the peninsula (in principle such as shown in Fig. 5.40).

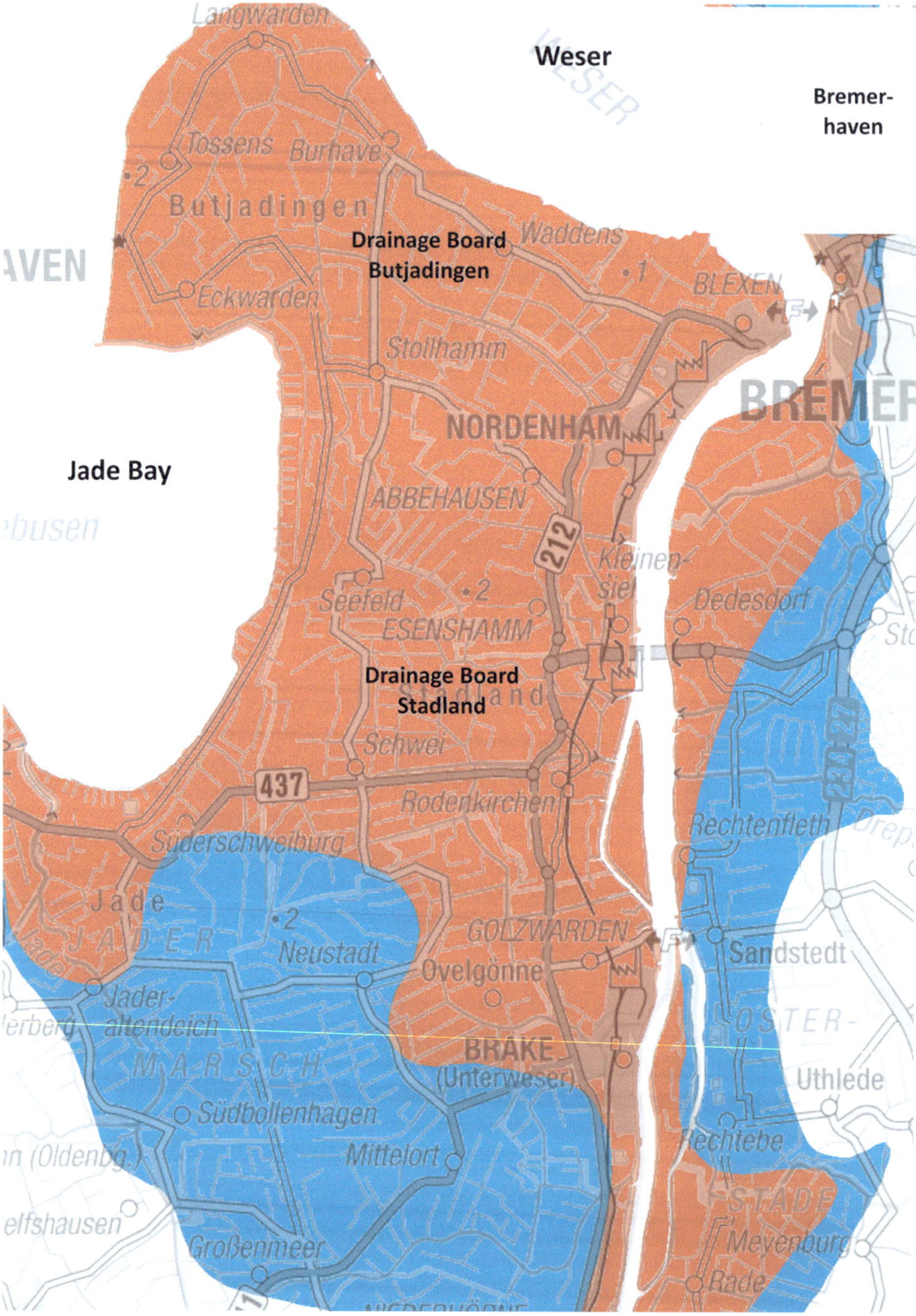

Fig. 5.40 Saline ground water body at the Weser estuary. The saline ground water body is penetrating underneath the northern part of the county Wesermarsch (*orange*: upper saline ground water body, *blue*: lower saline ground water body). Source: NIBIS® Kartenserver (2014): Versalzung des Grundwassers 1:200 000 (HÜK200) on nibis.lbeg.de

Habitat [ha]	End of nineteenth/early twentieth century*	Mid twentieth century**	Recent***	Difference [ha]
Elbe				
Subtidal	45,801	45,915	42,293	-3,508
Tidal flats	34,300	25,771	33,000	-1,300
Marsh	21,843	18,736	8,882	-12,961
Total Elbe				-17,769
Sea Schelde				
Subtidal	3,347	3,241	3,329	-18
Tidal flats	929	889	824	-105
Marsh	3,241	1,483	682	-2,559
Total Sea Schelde				-2,682
Weser				
Subtidal	56,917	52,995	54,088	-2,829
Tidal flats	40,764	43,255	40,322	-442
Marsh	7,609	4,793	6,306	-1,303
Total Weser				-4,574

Fig. 5.41 Habitat loss at the estuaries Schelde, Weser and Elbe. * Data covers different time spans from the end of the nineteenth century to the beginning of the twentieth century. Data for the Weser not available. ** Data covers different time spans for each estuary. *** Data for the Elbe from 1992–1995, for the Schelde 2000 and for the Weser 2005–2008. Source: TIDE Project Consortium (2013)

The orange area illustrates that the upper ground water body is saline and the blue color indicates that the lower saline ground water body is penetrating further inland. Therefore, the installation of wells to extract ground water is impossible. The width of inland intrusion of the salt water zone is depending on the ground water recharge at the higher areas and, for example, from the tide regime and soil properties (see Chap. 4). The fine-grained material of the marshes (former salt marshes) almost prevents ground water recharge in a reasonable amount.

Ecological Consequences in Brief

The change of tidal range indicates that considerable alterations have taken place in the estuary. These changes influence the ecological, chemical and physical composition of the estuary. A lot of habitats were destroyed or are under deterioration because of increasing current velocity (Fig. 5.41). Shallow water areas are either lost because of reclamation or more frequently inundated because of changes in high and low water regime. The brackish water zone is penetrating further upstream with consequences for flora and fauna, and in low-lying marsh lands also for water extraction for feeding cattle with fresh water.

Exercise

1. According to estuarine development the main driver for change has been identified as trade and shipping. The consequences of the change have been outlined. The interrelationship with natural dynamics and the disturbance of the natural environment was

very briefly touched. Elaborate on this issue of consequences for the development of the natural environment. What about the consequences for fishing and leisure crafts?
2. What are the interlinkages between hydrodynamics and morphological development of an estuary?
3. Develop alternative options in order to reduce the pressure on the estuaries imposed by shipping.

References

M. L. Allemeyer. *"Kein Land ohne Deich . . . !" Lebenswelten einer Küstengesellschaft in der Frühen Neuzeit*. Vandenhoek & Ruprecht, 2006.

M. L. Allemeyer. "...dass man dem grausam Toben des Meeres nicht etwa kann Widerstand thun mit Gewalt". Kontroversen um den Küstenschutz im 17. und 18. Jahrhundert. In N. Fischer, S. Müller-Wusterwitz, and B. Schmidt-Lauber, editors, *Inszenierungen der Küste*, pages 87–105. Reimer Verlag, Berlin, 2007.

Anonymous. The Venice system for the classification of marine water according to salinity. *Limnology and Oceanography*, 3:346–347, 1958.

O. Auhagen. Zur Kenntnis der Marschwirtschaft. Zwei Abhandlungen. *Landwirtschaftliche Jahrbücher*, 25:619–874, 1896.

C. Baeteman, M. Waller, and P. Kiden. Reconstructing middle to late Holocene sea-level change: methodological review with particular reference to 'A new Holocene sea-level curve for the southern North Sea presented by K.-E. Behre. *Boreas*, 40:557–572, 2011.

A. Bantelmann. Die Landschaftsentwicklung an der schleswig-holsteinischen Westküste, dargestellt am Beispiel Nordfriesland. Eine Funktionschronik durch fünf Jahrtausende. *Die Küste*, 14:5–99, 1966.

Bartels. Eigentliche Beschreibung der vor dieser Grafschaft zur See hinaus belegener Eylanden mit anheffter Erinnerung. *Jahrbuch der Gesellschaft für bildende Kunst und vaterländische Altertümer zu Emden*, 4:35–42, 1881.

K.-E. Behre. Eine neue Meeresspiegelkurve für die südliche Nordsee - Transgressionen und Regressionen in den letzten 10.000 Jahren. *Probleme der Küstenforschung*, 28:9–63, 2003.

K.-E. Behre. A new Holocene sea-level curve for the southern North Sea. *Boreas*, 36:82–102, 2007.

K.-E. Behre. *Landschaftsgeschichte Norddeutschlands*. Wachholtz Verlag, 2008.

M. Bergemann. Die Lage der oberen Brackwassergrenze im Elbeästuar. *Deutsche Gewässerkundliche Mitteilungen*, 39(4/5):134–137, 1995.

M. Bergemann. Die Trübungszone in der Tideelbe - Beschreibung der räumlichen und zeitlichen Entwicklung. Technical report, Wassergütestelle Elbe, Hamburg, 2004.

J. C. Besteman. Frisian salt and the problem of salt-making in North Holland during the Carolingian period. *Berichten van de Rijksdienst voor Oudheidkundig Bodemonderzoek*, 24:171–174, 1974.

Bezirksregierung Weser-Ems. Generalplan Küstenschutz für den Regierungsbezirk Weser-Ems. Technical report, Bezirksregierung Weser-Ems, 1997.

BioConsult. Comparative analysis of sediment management strategies in the estuaries of Humber, Scheldt, Weser and Elbe - Study in the framework of the Interreg IVB project TIDE. Technical report, NLWKN, 2013.

D. Blackbourn. *The Conquest of Nature. Water, Landscape, and the Making of Modern Germany*. W. W. Norton & Company, 2006.

D.P. Blok. Wie alt sind die ältesten niederländischen Deiche? Die Aussagen der frühesten schriftlichen Quellen. *Probleme der Küstenforschung*, 15:1–8, 1984.

A. Brahms. *Anfangs-Gründe der Deich- und Wasserbaukunst. 1. und 2. Teil. Nachdruck*. Marschenrat zur Förderung der Forschung im Küstengebiet, 1754.

K. Brandt. Die mittelalterliche Siedlungsentwicklung in der Marsch von Butjadingen (Landkreis Wesermarsch). *Siedlungsforschung. Archäologie-Geschichte-Geographie*, 2:123–146, 1984.

K. Brandt. Besiedlungsgeschichte der Nord- und Ostseeküste bis zum Beginn des Deichbaues. In J. Kramer and H. Rohde, editors, *Historischer Küstenschutz*, pages 17–38. Verlag Konrad Wittwer, Stuttgart, 1992.

F. Breuel. *Geschichte des Anwachsrechts in Ostfriesland*. Vandenhoeck und Ruprecht, 1954.

I. Coen. De eeuwige Schelde? Ontstaan en ontwikkeling van de Schelde. Technical Report D/2007/3241/203, waterbouwkundig laboratorium, 2008.

M.J. Collins and D. Miller. Upper Hudson river estuary (USA) floodplain change over the 20th century. *River Research and Applications*, page 8, 2011. doi: 10.1002/rra.1509.

N.D. Cutts and K.L. Hemingway. Analysis of the TIDE Estuarine Conflict Matrices. Technical report, University of Hull, Hull, 2013. URL www.tide-toolbox.eu.

[Departmental Committee on Coastal Flooding] DCCF. Die Sturmflut vom 31. Januar/1.Februar 1953 in England. *Die Küste*, 3:70–77, 1955.

J.G. de Ronde, D. Dilling, and M.E. Phillipart. Design criteria along the Dutch coast. *Proc. of HYDROCOAST '95*, pages 138–151, 1995.

Delta Commission. Eindverslag en Interimadviezen. Deel 1. Technical report, Deltaciommissie, 1961.

L. Denys and C. Baeteman. Holocene evolution of relative sea level and local mean high water spring tides in Belgium - A first assessment. *Marine Geology*, 124:1–19, 1995.

Deutscher Bundestag. Rahmenplan der Gemeinschaftsaufgabe Verbesserung der Agrarstruktur und des Küstenschutzes" für den Zeitraum 2007 bis 2010. Drucksache 16/5324, 2007a.

Deutscher Bundestag. Grundgesetz für die Bundesrepublik Deutschland, 2007b.

K.S. Dijkema. Geography of salt marshes in europe. *Z. Geomorph. N.F.*, 31:489–499, 1987.

P. Doody, M. Ferreira, S. Lombardo, I. Lucius, R. Misdorp, H. Niesing, A. Salman, and M. Smallegange. Living with Coastal Erosion in Europe. Sediment and Space for Sustainability. Results from the EUROSION Study. Technical report, European Commission, 2004.

J.-P. Ducrotoy. Ecological restoration of tidal estuaries in North Western Europe: an adaptive strategy to multi-scale changes. *Plankton Benthos Research*, 5:174–184, 2010.

[Advisory Committee for Coastal Protection] EAK. EAK 1993. Empfehlungen für die Ausführung von Küstenschutzwerken durch den Ausschuss für Küstenschutzwerke. *Die Küste*, 55:541, 1993.

[Advisory Committee for Coastal Protection] EAK. EAK 2002. Korrigierte Ausgabe 2007. Empfehlungen für die Ausführung von Küstenschutzwerken durch den Ausschuss für Küstenschutzwerke. *Die Küste*, 65:589, 2002.

[European Environment Agency] EEA. The changing face of Europe's coastal areas. Technical Report 6, EEA, 2006.

EMOVE Project Consortium. Work Package 3 - European Joint Fact Finding. Part 1: Problem Identification. Technical report, RWS, MOW, BAW, COWI, Middelburg, 2015.

[Engineering Committee for North and Baltic Sea] Engineering Committee. Die Sturmflut vom 16./17. Februar 1962 im niedersächsischen Küstengebiet. *Die Küste*, 10(1):17–54, 1962a.

[Engineering Committee for North and Baltic Sea] Engineering Committee. Empfehlungen für den Deichschutz nach der Februar-Strumflut 1962. *Die Küste*, 10(1):113–130, 1962b.

L. Franzius. Die Korrektion der Unterweser (Nachdruck von 1888). *Die Küste*, 51:39–74, 1888/1991.

J. Fries-Knoblach. Gerätschaften, verfahren und bedeutung der eisenzeitlichen salzsiederei in mittel- und nordwesteuropa. Techical report, Leipziger Forschungen zur Ur- und Frühgeschichtlichen Archäologie, 2001.

L. Geerts, K. Wolfstein, S. Jacobs, S. van Damme, and W. Vandenbruwaene. Zonation of the TIDE estuaries. Technical report, Antwerp Port Authority (APA), Hamburg Port Authority (HPA), 2012. URL www.tide-toolbox.eu.

G. Gönnert, S.K. Dube, T. Murty, and W. Siefert. Global Storm Surges. *Die Küste*, 63:623, 2001.

K.J. Gregory. The human role in changing river channels. *Geomorphology*, 79(3-4):172–191, 2006.

D. Hallewas. Mittelalterliche Seedeiche im holländischen Küstengebiet. *Probleme der Küstenforschung*, 15:9–27, 1984.

R. Hillen and T. de Haan. Development and implementation of the coastal defence policy for The Netherlands. In R. Hillen and H.J. Verhagen, editor, *Coastline of the Southern North Sea*, pages 188–201. ASCE, Louisiana, 1993.

J.L.A. Hofstede and R. Schirmacher. Volandmanagement in Schleswig-Holstein. *Die Küste*, 58: 61–73, 1996.

H. Homeier. Das Wurster Watt – Eine historisch-morphologische Untersuchung des Küsten- und Wattgebietes von der Weser- bis zur Elbmündung. In *Jahresbericht 19/1967*, pages 1–121. Forschungsstelle Norderney, 1969.

H. Homeier, H.-J. Stephan, and H.D. Niemeyer. Historisches Kartenwerk Niedersächsische Küste der Forschungsstelle Küste. In *Bericht FSK 43/2010*, pages 1–133. Forschungsstelle Küste Norderney, 2010.

[Hamburg Port Authority and Wasser- und Schifffahrtsverwaltung] HPA and WSV. Work Package 5 Measures - Dredging and Disposal Strategies. River Elbe. Technical report, HPA and WSV, 2012.

IAHR. Scheldt Estuary - Physics and integrated management. Technical report, Rijkswaterstaat, 2015.

S. Jelgersma. Sea-level changes in the North Sea basin. In E. Oele, R.T.E. Schüttenhelm, and J.A. Wiggers, editors, *The Quaternary History of the North Sea*, pages 233–248. Acta Univ. Ups. Symp. Univ. Annum Quingentesium Celebrantis 2, Uppsala, 1979.

O. Knotterus. History of human settlement, cultural change and interference with the marine environment. *Helgol. Marine Research*, 59:2–8, 2005.

I. Krämer, C. Hein, and J. Janssen. Auswirkungen übergroßer Containerschiffe (ULCS). In PIANC Germany, editor, *XXXII. Internationaler Schifffahrtskongress*, pages 6–12, Bonn, 2010. BMVBS.

J. Kramer and H. Rohde. *Historischer Küstenschutz. Deichbau, Inselschutz und Binnenentwässerung and Nord- und Ostsee*. Verlag Konrad Wittwer, Stuttgart, 1992.

H. Kunz. *Wasserhaushaltsuntersuchungen in tidebeeinflussten Gebieten*. Dissertation, Technical University Hannover, 1975.

H. Kunz. Salinity and water levels in the Weser estuary during the last hundred years – Anthropogenic influences on the coastal environment. In *Coastal Engineering*, pages 3533–3547, Kobe, 1995. ASCE.

H. Kunz. Küstenschutz- und Küstenzonenmanagement: Gesamtschau einer Dokumentation. In *Küstenschutz- und Küstenzonenmanagement - eine Dokumentation*, Arbeiten aus der Forschungsstelle Küste, Heft 15, pages 13–78. Niedersächsisches Landesamt für Ökologie, 2004a.

H. Kunz. Sicherheitsphilosophie für den Küstenschutz. *Jahrbuch der Hafenbautechnischen Gesellschaft*, 54:253–287, 2004b.

J.M. Lorenzen. Hundert Jahre Küstenschutz an der Nordsee. *Die Küste*, 3(1/2):18–32, 1955.

J.M. Lorenzen. Über Aufgaben und Organisation des Küstenausschusses Nord- und Ostsee. *Die Küste*, 14(2):1–4, 1966.

T. Lucker, D. Busch, J. Knötzel, and M. Schirmer. Unterweserausbau, Entwicklung des Schiffsverkehrs und Auswirkungen auf das Flusssystem. In B. Gerken and M. Schirmer, editors, *Die Weser. Limnologie aktuell*, volume 6, pages 301–312, Bremen, 1995.

K. H. Marschalleck. Die Salzgewinnung an der friesischen Nordsseeküste. *Probleme der Küstenforschung im südlichen Nordseegebiet*, 10:127–150, 1973.

D. Meier. *Die Nordseeküste*. Boyens Verlag, 2006.

D. Meier. *Die Unterelbe*. Boyens Verlag, 2014.

[Ministry for Food, Agriculture and Forestry and the Senator for Infrastructure Bremen] MFAF and SIB. Sturmfutschutz durch Sperrwerke. Technical report, Ministry for Food, Agriculture and Forestry and the Senator for Infrastructure Bremen, 1979.

[Ministerie van Infrastuctuur en Milieu and Ministerie van Economische Zaken] MinInfraM and MinEZ. Delta Programme 2016. Working on the delta. And now we're starting for real. Technical Report, MinInfraM and MinEZ, 2015.

[Ministry of Transport, Public Works and Water Management] MVenW. A new coastal defence policy for The Netherlands. Technical report, Ministry of Transport, Public Works and Water Management, 1990.

[Ministry of Transport, Public Works and Water Management] MVenW. Naar integraal kustzonebeleid. Beleidsagenda voor de kust. Technical report, Ministry of Transport, Public Works and Water Management, 2002.

[Ministry of Transport, Public Works and Water Management] MVenW. Floris study - Full Report. DWW-2006-014, Ministry of Transport, Public Works and Water Management, 2005a.

[Ministry of Transport, Public Works and Water Management] MVenW. Veiligheid Nederland in Kaart. Tussenstand onderzoek overstromingsrisico's. DWW-2005-074, Ministry of Transport, Public Works and Water Management, 2005b.

[Ministry of Transport, Public Works and Water Management] MVenW. Veiligheid Nederland in Kaart. Hoofdrapport onderzoek overstromingsrisico's. DWW-2005-081, Ministry of Transport, Public Works and Water Management, 2005c.

[Ministry of Transport, Public Works and Water Management] MVenW. Water Act. Technical Report, MVenW, 2010.

NDG. Niedersächsisches Deichgesetz – State Law on Dikes for Lower Saxony. Nds. GVBl. 6/2004, Federal Ministry of the Environment (Lower Saxony), 1963.

[Niedersächsisches Ministerium für Ernährung, Landwirtschaft und Forsten] Nds. MELF. Generalplan Küstenschutz Niedersachsen. Technical report, Niedersächsisches Ministerium für Ernährung, Landwirtschaft und Forsten, 1973.

P. H. Nienhuis. *Environmental History of the Rhine-Meuse Delta: An ecological story on evolving human-environmental relations coping with climate change and sea-level rise*. Springer Netherlands, Dordrecht, 2008. ISBN 978-1-4020-8213-9.

[Niedersächsischer Landesbetrieb für Wasserwirtschaft und Küstenschutz – Betriebsstelle Norden] NLWK Norden. Vorlandmanagementplan für den bereich der deichacht norden. Technical report, NLWK Norden, 2003.

[Niedersächsischer Landesbetrieb für Wasserwirtschaft, Küsten- und Naturschutz] NLWKN. Generalplan Küstenschutz Niedersachsen/Bremen – Festland. Technical report, Niedersächsischer Landesbetrieb für Wasserwirtschaft, Küsten- und Naturschutz, 2007.

NLWKN and BioConsult. Sediment Management Strategies in the Weser Estuary - Study in the framework of the Interreg IVB project TIDE. Technical report, NLWKN, 2012.

OECD. OECD Environmental Performance Reviews. A practical introduction. Technical Report OECD/GD(97)35, OECD, 1997.

C. H. Ovesen. *Proceedings of the Second Trilateral Working Conference on Saltmarsh Management in the Wadden Sea Region*. Ministry of the Environment, 1989.

E.B. Peerbolte. *Sea-level Rise and Safety. A consideration of safety impacts in low-lying coastal areas with particular reference to the Netherlands*. PhD thesis, University of Twente, 1993.

N. Pirrone, G. Trombino, S. Cinnirella, A. Algieri, G. Bendoricchio, and L. Palmeri. The Driver-Pressure-State-Impact-Response (DPSIR) approach for integrated catchment-coastal zone management: preliminary application to the Po catchment-Adriatic Sea coastal zone system. *Reg Environmental Change*, 5(2-3):111–137, 2005.

W. Reinhardt. Zum frühen Deichbau im niedersächsischen Küstengebiet. *Probleme der Küstenforschung*, 15:29–40, 1984.

[Rijkswaterstaat] RWS. Water Management in the Netherlands. Technical Report, RWS, 2011.

SafeCoast. Coastal flood risk and trends for the future in the north sea region. results and recommednation of the project safecoast. Synthesis Report, EU Interreg IIIB North Sea Programme, SafeCoast Project Consortium, 2008.

J.W. Seijffert. Coastal Defence in the Netherlands. In *Strategien des Küstenschutzes. Tagungsband zur Veranstaltung am 21. September 2001 in Hannover*, Kolloquien-Reihe. Bilanz und Perspektiven der Umweltpolitik. Niedersächsisches Umweltministerium, 2001.

A. Siegmüller and F. Bungenstock. Salztorfabbau im Jadebusengebiet. Prospektion von anthrogenen Landabsenkungen und ihren Folgen. *Nachrichten aus Niedersachsens Urgeschichte*, 79:201–220, 2010.

K. Slager. *De ramp: een reconstructie*. Goes: De Koperen Tuin, 1992.

E. Smeets and R. Weterings. Environmental indicators: Typology and overview. Technical Report Technical Report No. 25, EEA - European Environment Agency, 1999.

M. Stock, K. Kiehl, and H.D. Reinke. Salzwiesenschutz im Schleswig-Holsteinischen Wattenmeer: Grundlagen, Zielsetzung und bisherige Umsetzung. Technical report, UBA-FB 93-101, 1994.

H. Streif. *Coastal Lowlands: Geology and Geotechnology*, chapter Barrier islands, tidal flats, and coastal marshes resulting from a relative rise of sea level in East Frisia on the German North Sea coast, pages 213–223. Springer Netherlands, Dordrecht, 1989. ISBN 978-94-017-1064-0.

H. Streif. Geologische Aspekte der Klimawirkungsforschung im Küstenraum der südlichen Nordsee. In H.-J. Schellnhuber and H. Sterr, editors, *Klimaänderung und Küste. Einblick ins Treibhaus*, pages 77–93. Springer Verlag, 1993.

[Technical Advisory Committee on Water Defence] TAW. Fundamentals on Water Defence. Technical report, TAW, 1998.

[Technical Advisory Committee on Water Defence] TAW. From Probability of Exceedance to Probability of Flooding. Technical report, TAW, 2000.

O. Tenge. *Der Jeversche Deichband: Geschichte und Beschreibung der Deiche, Uferwerke und Siele im Dritten Oldenburgischen Deichbande und im Königlichen Preußischen westlichen Jadegebiet*. III. Oldenburgischer Deichband, 1898.

O. Tenge. *Der Butjadinger Deichband: Geschichte und Beschreibung der Deiche, Uferwerke und Siele im Zweiten Oldenburgischen Deichbande und im Königlichen Preußischen östlichen Jadegebiet*. II. Oldenburgischer Deichband, 1912.

TIDE Project Consortium. Management of Estuaries. The need to understand nature and society. TIDE summary report. Technical report, Hamburg Port Authority (HPA), Hamburg, 2013. URL www.tide-toolbox.eu.

G. Tomczak. Was lehrt uns die Holland-Sturmflut 1953? *Die Küste*, 3(1/2):78–95, 1955.

D. van Danzig. Economic Decision Problems for Flood Prevention. *Econometrica, Journal of the Econometric Society*, 24(3):276–287, 1956.

O. van de Plassche and T.B. Roep. Sea-level changes in the Netherlands during the last 6500 years: basal peat vs. coastal barrier delta. In D.B. Scott, P.A. Pirazzoli, and C.A. Honig, editors, *Late Quaternary Sea-level Correlation and Applications*, pages 41–56. Kluwer Academic Publishers, 1989.

G.P. van de Ven. *Man-made lowlands, history of water management and land reclamation in The Netherlands*. ICID and Royal Institute of Engineers in the Netherlands, 1993.

W.E. van Duin, K.S. Dijkema, J. Bunje, T.F. Pedersen, and M. Stock. Salt marshes. In F. de Jong, J. Bakker, K. van Berkel, N. Dankers, K. Dahl, C. Gätje, H. Marencic and P. Potel, editor, *Wadden Sea Quality Status Report 1999*, volume 9, pages 165–170. Common Wadden Sea Secretariat, Wilhelmshaven, 1999.

R. Vergouwe. The National Flood Risk Analysis for The Netherlands. Technical report, Ministrie van Vekeer en Waterstaat, Interprovincial Overleg, Dutch Water Authorities, 2015.

A. Vierlingh. *Tractaet van Dyckagie*. Martinus Nijhoff, Den Haag, 1920.

J. von Gierke. *Die Geschichte des deutschen Deichrechts. Teil I und II.* Scienta Verlag Aalen, 1917.

M. Voorendt. The development of the Dutch flood safety strategy. Technical Report, TU Delft, 2015.

J. Vroon, C. Storm, and J. Coosen. Westerschelde, stram of struis? Eindrapport van het Project Oostwest, een studie naar beeinvloeding van fysische en verwante biologische partonen in een estuarium. Technical report, RIKZ, 1997.

V. Wetzel. Der Ausbau des Weserfahrwassers von 1921 bis heute. *HTG Jahrbuch der Hafenbautechnischen Gesellschaft*, 43:83–105, 1988.

C. Wöbcken. *Deiche und Sturmfluten*. Neudruck Walluf bei Wiesbaden, 1924.

C. Wöbcken. *Das Land der Friesen und seine Geschichte*. Schulzesche Verlagsbuchhandlung, 1932.

D. Yska. *Van Deltacommissie tot Deltacommissie. De rol van adviescommissies in de besluitvorming over veligheidsnormen voor hoogwaterbescherming*. Master Thesis, University Twente, 2009.

6 Prospects: Approaches of Integrated Management

6.1 Integrated Water Management for Quality and Quantity

6.1.1 European Water Framework Directive

After a long consultation process the Water Framework Directive (WFD, EC 2000) was published in 2000 as legislation for an integrated approach on water quality aspects, that is, ecological and chemical status of water bodies in the European Union. Hence, the WFD is the most important European legislations with regard to the context of this book. Various other Directives also regulate the water sector such as the Groundwater Directive (first 80/68/EEC, currently 2006/118/EC) and the Drinking Water Directive (98/83/EC) which are altogether subsumed as daughter Directives of the WFD.

European water legislation[1] began with setting standards for rivers and lakes used for drinking water abstraction in 1975, and led in 1980 to quality targets for drinking water. In 1988 the existing legislation was reviewed and a number of improvements and gaps were identified. This resulted in 1991 in the adoption of, for example, the Urban Waste Water Treatment Directive (91/271/EEC), providing a more stringent treatment of waste water where necessary and the Nitrates Directive (91/676/EEC), addressing water pollution by nitrates from agriculture.

Further results were proposals on a new Drinking Water Directive (98/83/EC), reviewing the quality standards and, where necessary, tightening them and a Directive for Integrated Pollution and Prevention Control (IPPC, 96/61/EC), adopted in 1996 and substantially amended in 2008 (2008/1/EC), addressing pollution from large industrial installations which was replaced by the Directive on Industry Emissions (2010/75/EC) in 2010.

[1] See ec.europa.eu/environment/water/water-framework/

F. Ahlhorn, *Integrated Coastal Zone Management*,
Wasser: Ökologie und Bewirtschaftung, DOI 10.1007/978-3-658-17052-3_6

Proposals for a general review of the Community water policy arose in middle of the 1990s. The European Water Policy should address the increasing awareness for water and water management, that is, to develop a coherent policy on European level. To achieve this, the European Water Policy was developed in an open consultation process involving all interested parties.

In 1996 a Water Conference constituted the end of this consultation process. This Conference was attended by representatives of Member States, regional and local authorities, water providers, industry, agriculture and consumers as well as environmentalists. The outcome of the consultation process was a broad consensus on the fragmentation of the water policy, in terms both of objectives and of means. All attendees agreed on the need for a single piece of legislation to resolve these problems. In response to this, the Commission presented a Proposal for a Water Framework Directive with the following key aims:

- expanding the scope of water protection to all water bodies, that is surface waters and ground water,
- achieving "good status" for all water bodies by a fixed deadline,
- water management based on river basins,
- intense public participation,
- streamlining legislation.

Finally, the Water Framework Directive (200/60/EC) was initiated with the intention to establish a framework for the protection of inland surface, transitional, coastal and ground water (Article 1 EC, 2000).

It should prevent deterioration, improve and enhance the status of aquatic and terrestrial ecosystems as well as wetlands directly depending on the aquatic ecosystem to promote sustainable water use, ensure the progressive reduction of pollution of ground water (Article 1 paragraphs a to e). In Article 2 the definition for coastal waters is provided (paragraph 7): *coastal waters* means surface water on the landward side of a line, every point of which is at a distance of one nautical mile on the seaward side from the nearest point of the baseline from which the breadth of territorial waters is measured, extending where appropriate up to the outer limit of transitional waters.

According to the WFD the focus is on the management of water bodies at river basin scale – the natural geographical and hydrological unit – instead of administrative or political boundaries. Actions taken by the Member States concerned for the Maas, Schelde or Rhine river basins have served as positive examples of this approach, with their cooperation and joint objective-setting across Member State borders, in Germany followed by processes at the rivers Ems and Elbe.

There are a number of objectives to protect the quality of water bodies. The key ones are general protection of the aquatic ecology, specific protection of unique and valuable habitats, protection of drinking water resources, and protection of bathing water. All these objectives must be integrated for each river basin. Ecological protection should apply to all water bodies with the central aim that the environment is protected to a high level in its entirety.

The Directive also provides a framework for integrated management of ground water and surface water for the first time at European level. For this reason the implementation of the WFD is aiming at the achievement of a *good ecological status* and a *good chemical status*.

Surface Water

Ecological Protection

The good ecological status is defined in Annex V of the WFD. It comprises the quality of the biological community, the hydrological characteristics and the chemical characteristics. Because of ecological variability no absolute standards for biological quality can be set which apply across the Community.

Chemical Protection

Good chemical status is defined in terms of compliance with quality standards established for chemical substances at European level. The Directive provides a mechanism for renewing these standards and establishing new ones by means of a prioritization mechanism for hazardous chemicals.

Other Uses

The other uses or objectives for which water is protected apply only in specific areas. Therefore, the obvious way to integrate them is to designate specific protection zones to meet different objectives. The overall plan of objectives for the river basin will then require ecological and chemical protection everywhere as a minimum, but where more stringent requirements are needed for particular uses, zones will be established and higher objectives set within them.

A set of uses which adversely affect the status of water but which are considered essential on their own terms are, for example, flood protection (see Flood Risk Management Directive on p. 139) and drinking water supply. Less apparent cases are navigation and power generation, where the activity is open to alternative approaches (transport can be switched to land, other means of power generation can be used). Interferences may be acceptable, but have to pass three tests: that the alternatives are technically impossible, that they are prohibitively expensive or that they produce a worse overall environmental result.

Ground Water

Chemical Status

The general target to ground water protection is that it should not be polluted at all. For this reason, setting chemical quality standards may not be the best approach, as it gives the impression of an allowed level of pollution. A few standards have been established at European level for particular issues (nitrates, pesticides and biocides). The general protection comprises a prohibition on direct discharges into the ground water body, and a requirement to monitor ground water bodies so as to detect changes in chemical composition.

Quantitative Status

Quantity is a major issue for ground water. There is only a certain amount of recharge into ground water each year, and of this recharge, some is needed to support connected ecosystems. For good management, only that portion of the overall recharge not needed by the ecology can be abstracted. The Directive limits ground water abstraction either for drinking water or irrigation purposes to that quantity.

Pollution Control

In the past the approach to pollution control at European level was concentrating on the source mainly by technological approaches. Other approaches on pollution control were dealing with the capacity of the receiving environment in the form of quality objectives.

Due to methodological problems in applying one of these objectives, a consensus was developed that both approaches are needed in practice. Regarding the source, all existing technology-driven and source-based control measures must be implemented as a first step (e.g., polluter-pays-principle). The framework comprises a list of priority substances to take action at EU level, prioritized on the basis of risk. Afterwards, the design of the most cost-effective set of measures to achieve the reduction of the load of those substances is required, taking into account both product and process sources. If measures taken on the source side are not sufficient to achieve these objectives it urges responsible organizations to take additional measures.

The River Basin Management Plan

The river basin management plan is a detailed document of how the objectives set for the river basin (ecological status, quantitative status, chemical status and protected area objectives) shall be achieved. This plan has to include all results of a comprehensive analysis: the river basin's characteristics, a review of the impact of human activity on the status of water bodies in the basin, the remaining gap for meeting these objectives and a set of measures designed to fill the gap. An additional component is that an economic appraisal of water use within the river basin must be carried out to provide the basis for a rational discussion on the cost-effectiveness of possible measures. Public participation is essential for the preparation of river basin management plans.

Public Participation

There are two main reasons for intense public participation. The first is that the decisions on the most appropriate measures to achieve the objectives in the river basin management plan will be the need to balance the interests of various groups. An economic analysis might provide a rational basis for this, but it is essential that the process is open to the scrutiny of those who will be affected.

The second reason concerns enforceability. The greater the transparency in the establishment of objectives, the implementation of measures, and the reporting of standards by responsible governmental bodies the greater the support of all affected or at least interested parties.

The WFD is concentrating on the quality of the Community water bodies (Fig. 6.1). We have seen that not only the quality is a problem in coastal areas also the quantity seems to be problematic, that is, storm surges and fluvial floods. The WFD does not deal with these issues in an explicit way, thus, an integrated policy dealing with these issues was missing since 2000. In 2007 the European Flood Risk Management Directive (FRMD) went into force dealing explicitly with this gap. But before the European FRMD will be introduced an excursus is made to the international development in dealing with flood risks and the explanation of the term risk.

6.1.2 Transition to Flood Risk Management

Introduction

As reaction to the results of the World Commission on Environment and Development (WCED, 1987) in 1987 the Intergovernmental Panel on Climate Change (IPCC) was installed in 1988 by the United Nations (UN) and the World Meteorological Organization (WMO). The Panel is organized in working groups with several subgroups focusing on specific issues. Today, three Working Groups are established: Working Group I assesses the physical scientific aspects of the climate system and climate change, Working Group II assesses the vulnerability of socio-economic and natural system to climate change and, finally, Working Group III assesses options for mitigating climate change.

Fig. 6.1 River Jade in Lower Saxony, Germany. ©Frank Ahlhorn

For the First Assessment Report (FAR) in 1990 the formerly Working Group III was focusing on response strategies to climate change. A subgroup for Coastal Zone Management was installed to work on options and adaptation to sea level rise and other impacts of global change on coastal areas (IPCC, 1990a). The Second Assessment Report (SAR) published in 1995 contained the contribution of the subgroup Coastal Zone Management in the official Working Group II which was charged with providing information and recommendations to national and international policy on coastal zone management strategies and long-term policies on adapting to climate change and sea level rise – see IPCC (1996).

Currently, the Fifth Assessment Report (AR5) was published in 2013 and 2014 – see IPCC (2013, 2014a,b,c). Hereafter, adaptation strategies for coastal zones will be described as compiled and shortly described in the beginning of the IPCC.

Adaptive Strategies in Coastal Zones

In 1989 a group of experts met in Florida to discuss adaptive options and policy implications of sea level rise (IPCC, 1989). The experts conducted a problem identification of sea level rise on different types of coastal zones and compiled possible adaptation strategies. Finally, four possible strategies where identified (IPCC, 1989, p. 3–4):

- *defending* a site to maintain its existing uses,
- *adapting in place* by modifying structures and various activities to accommodate rising seas,
- *retreating landward*, spending resources on relocation rather than on coastal defenses,
- *employing temporary solutions* until escalating economic, social, and resource costs require a different approach, at which time on of the previous options can be implemented.

These experts categorize the potential response options which could be applied to achieve one of the previous strategies in (see IPCC, 1989, p. 4):

- *Technical, engineering and structural*: These responses include the construction of infrastructure and also soft measures such as dikes, barriers, sand nourishment or raising of coastal land by filling. On the one hand, it was acknowledged that some of these responses are very costly and could result in significant environmental impacts. On the other hand, they emphasized that these responses could be extremely effective in protecting existing land and structures and that they were in several countries already well established and continuously improved.
- *Natural, biological and ecological*: These options can mitigate the impacts of rising sea level by replacing lost resources or by developing alternative habitat areas that could serve similar ecological functions. Options include creating wetlands and dunes, stabilizing dunes by planting vegetation, and planting mangroves.

- *Nonstructural*: The simplest approach is to allow coastal resources and land uses to naturally respond to the changing conditions. If complete inaction is unacceptable, non-structural options can help reduce the risk to property and the environment by removing structures and directing populations away from vulnerable areas. Resettlement can be encouraged by regulatory and legal measures that
  - require structures to be removed,
  - prohibit rebuilding of structures under special circumstances (e.g., after significant storm damage),
  - prohibit private construction of bulkheads,
  - establish restrictions on new development through zoning or other means in order to reduce population concentrations.

With regard to the third response option the experts explicate that "the process of gradual retreat from areas threatened by sea level rise may require new institutional arrangements to coordinate various levels of governmental decision making. It will also require public education to increase awareness for all sectors of society about both the impacts of sea level rise and the implications of various adaptive responses" (IPCC, 1989, p. 4). Subsequently, in the report of the Coastal Zone Management Subgroup (IPCC, 1990b) on strategies for adaptation to sea level rise the elaboration on the responses were introduced as *retreat*, *accommodate* and *protection* (Fig. 6.2).

Possible options in the category *retreat* were divided into three different classes: (i) preventing development, (ii) allowing temporarily limited development and (iii) no direct government role.

The first option is linked to the effort of limiting the development in vulnerable coastal zone entirely. Consequently, government has to take precautionary arrangements such as land acquisition, land use restriction or reduction of subsidies and incentives for the development in vulnerable coastal areas. In the second case development could take place in accordance with the rules set by the government which could encompass "(i) regulations that prohibit private construction of protective structures or (ii) conversion of land ownership to long term or conditional leases which expire when the sea reaches a particular level or when the property owner dies" (IPCC, 1990b, p. 6). According to the third option it is explicated that

> it would to be depend on the working of the private market. Productive crop and timber lands may be left to slowly and progressively deteriorate as a result of salt intrusion into the groundwater or by surface flooding. Wells and surface water exposed to salt water intrusion would gradually be abandoned [...] Under this option, governments could take the more limited role of ensuring that all participants in potentially vulnerable areas have full knowledge about the expected sea level rise and its associated uncertainties. Development presumably not occur if developers, lenders and insurers were not willing to accept the risks. However, if people continue to build in vulnerable areas, governments must be prepared to take the necessary actions to ensure public safety. (IPCC, 1990b, p. 6)

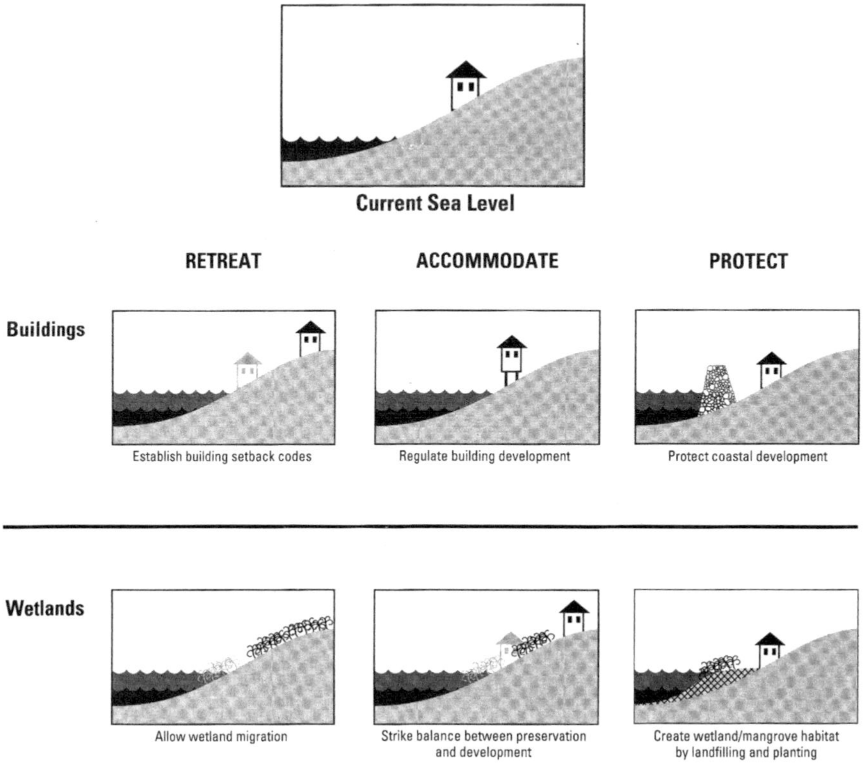

Fig. 6.2 Three adaptation categories are represented as identified by the IPCC subgroup CZM: retreat, accommodate and protect. The category *retreat* means abandonment of land structures in vulnerable areas and resettlement of inhabitants. The category *accommodation* is linked to continued occupancy and use of vulnerable areas. Finally, the category *protect* is connected to the defense of vulnerable areas, especially population centers, economic activities and natural resources. Source: adapted from IPCC (1992)

Accommodation

requires advanced planning and acceptance that some coastal zone values could be lost. Many coastal structures, particularly residential and small commercial buildings, could be elevated on pilings for protection from floods. To counter surging water and high winds, building codes could specify minimum floor elevations and piling depths, as well as structural bracing. Storm warning and preparedness plans could be instituted to protect the affected population from extreme events. Where saltwater damages agricultural lands and traditional crops, salt tolerant crops may be feasible alternative. Fundamental changes in land use may be desirable, such as the conversion of some agricultural lands into aqua-cultural uses.

Human activities that destroy the natural protection values of coastal resources can be prohibited. Perhaps the most important controls would be to prohibit filling wetlands, damming rivers, mining coral and beach sands and cutting mangroves. undeveloped land with sufficient elevation and slope can be set aside to accommodate natural reestablishment of wetlands and mangroves. Within deltaic areas, natural processes can be maintained by diverting water and sediment. In response to salinity intrusion into groundwater aquifers, management controls can be implemented to regulate pumping and withdrawal practices.

Requiring private insurance coverage in vulnerable areas is an important method to compensate injuries and damages caused by natural disasters. It forces people to consider whether risks are worth taking and provides the necessary funds to repair damages and compensate victims. (IPCC, 1990b, p. 7)

Protection

involves defensive measures and other activities to protect areas against inundation, tidal flooding, effects of waves on infrastructure, shore erosion, salinity intrusion and the loss of natural resources. The measures may be drawn from an array of *hard* and *soft* structural solutions. they can be applied alone or in combination, depending on the specific conditions of the site.

There is no single or generic best solution, as each situation must be evaluated and treated on its particular merits. However, there are some basic steps in the selection of measures likely to produce the highest economic returns. First, those charged with planning, design or management responsibilities in the coastal zones should be cognizant of the potential for future sea level rise. Moreover, proposed plans should leave options open for the most appropriate future response. For example, many protection structures can be planned and designed with features that allow for future incremental additions which, if needed, could accommodate increased water levels and wave action. This can often be done without significant additional costs in the initial investment. (IPCC, 1990b, p. 7)

Addressing (Flood) Risk Management

In Sect. 5.1.2 the development of coastal protection in the southern North Sea has been described for The Netherlands and Germany. Figure 5.4 outlines the development intersected and forced to make progress by disastrous events such as storm surges or river floods. The intention of this section is the transition from the safety approach in coastal protection to the flood risk management approach. Although the safety approach inherently contains uncertainties, for example, for the calculation of parameters to determine the crest height of a main dike (see 5.2), it does not fully comprise the flood risk management approach.

Within the European FRMD (p. 139) the intention of *flood risk management* is defined as "a framework for the assessment and management of flood risks, aiming at the reduction of adverse consequences for human health, the environment, cultural heritage and economic activity" (Article 1). Thus, the Directive and the term flood risk is intended to be valid for rivers, river basins and coastal waters.

Beforehand, some aspects concerning risk will be explained: here, the delineation is linked to storm surges.

Risk, Uncertainty and Vulnerability

The European Flood Risk Management project (FloodSite[2]) collated in its first phase information about the language of risk in different partner countries (see Gouldby and Samuels, 2005).This collation serves as basis for the explanation of terms concerning risk:

[2]www.floodsite.net, access: Apr. 2017

Risk is the combination of the probability of a consequence and its magnitude. Therefore risk considers the frequency or likelihood of occurrence of certain states or events (often termed "hazards") and the magnitude of the likely consequences associated with those exposed to these hazardous states or events.

Uncertainty exists where there is a lack of knowledge concerning outcomes. Uncertainty may result from an imprecise knowledge of the risk, that is, where the probabilities and magnitude of either the hazard and/or their associated consequences are uncertain. Even when there is a precise knowledge of these components there is still uncertainty because outcomes are determined probabilistically.

Risk analysis is a methodology to objectively determine risk by analyzing and combining probabilities and consequences.

Risk perception is the view of risk held by a person or group and reflects cultural and personal values, as well as experience.

Risk management is the complete process of risk analysis, risk assessment, option appraisal and implementation of risk management measures.

Risk communication is any intentional exchange of information on environmental and/or health risks between interested parties.

Vulnerability is the characteristic of a system that describes its potential to be harmed. This can be considered as a combination of susceptibility and value.

Ambiguity existed for the determination of risk, for which different formulas are available. In technical projects risk will be determined by the product of probability of failure and vulnerability (Gouldby and Samuels, 2005). For practical reasons and the purpose of this book the following definition will be adopted:

$$Risk = Probability\ of\ Failure \times Vulnerability \tag{6.1}$$

Risk can be tackled by influencing two items: the probability of failure or the vulnerability. The German Advisory Council on Global Change (WBGU) proposes a system of classifying risks (WBGU, 1999), which is based on the following criteria (p. 58):

- Probability of occurrence
- Magnitude of damage
- Ubiquity (spatial dimension of damage)
- Persistence (contaminants accumulated over long periods)
- Irreversibility (damage cannot be remedied)

- Delay effect (time period between the event and the damage)
- Mobilization (risks lead to severe conflicts and dread among the general public)

The WBGU differentiates between eight types of risk and suggests a decision tree to classify these types. Three areas were identified for the different types of risk: (a) normal, (b) transitional and (c) prohibited area (see Fig. 6.5 and see also Renn and Klinke 2013). The classification of these areas is determined by the probability of occurrence and the extent of the damage. *Normal* comprises risk of every day life which might be handled by simple measures. The risk area *transition* contains risks which can develop over time to one of the other areas and, therefore, either decrease or increase in intensity. Risk types assigned to the *prohibited* area are severe catastrophic events which show high extend of damage and low probability of occurrence. Markau (2003) concludes after the determination and evaluation of the risk of flooding, that storm surges are of the cyclops type and hence are located on the border between the transitional and prohibited area.

The cyclops risk type demands strategies as shown in Fig. 6.3. Three strategies have been identified to deal with the cyclops risk type: (a) ascertain the probability of occurrence, (b) prevent surprises and (c) emergency management (see, e.g., the Dutch approach of Multi-layered Safety in Sect. 6.1.4). Concerning the first strategy the need to determine or to assess the probability of occurrence based on research or technical measures is

Strategies	Tools
1. Ascertaining the probability of occurrence *P*	• Research to ascertain numerical probability *P* • International monitoring through – National risk centers – Institutional networking – International Risk Assessment Panel • Technological measures aimed at estimating probabilities
2. Preventing surprises	• Strict liability • Compulsory insurance for risk generators (e.g., floods, settlements) • Capacity building (licensing procedures, monitoring, training etc.) • Technological measures • International monitoring
3. Emergency management	• Human-resource and institutional capacity building (emergency prevention, preparedness and response) • Education, training, empowerment • Technological protective measures, including containment strategies • International emergency groups (e.g., fire services, radiation protection etc.)

Fig. 6.3 Strategies and instruments for the risk type cyclops. Source: WBGU (1999)

emphasized. Within the second strategy technical measures have to be taken to prevent surprises, and for the third it is recommended that the individual and institutional capacities be strengthened and technical measures to restrict the damages be taken.

Further essential component is the knowledge and evaluation of the vulnerability of coastal zones, which refers directly to the second term of the risk formula. For example, various aspects of vulnerability to climate change impacts are outlined in Klein and Nicholls (1999). An overview on state of the art for regional and local vulnerability assessment can be found in Sterr et al. (2000). The overview proposes an integrated process to identify the most appropriate coastal adaptation strategy. Sterr et al. (2000) evaluated two approaches to assess the vulnerability of coastal regions: the IPCC Common Methodology for Assessing the Vulnerability of Coastal Areas to Sea-Level Rise (result of the application for The Netherlands the ISOS-study by Peerbolte et al. 1991) and the methodology developed by UNEP which is based on the IPCC methodology. Several disadvantages have been identified, such as shortage of accurate and complete data or that the methodology is "less effective in assessing the wide range of technical, institutional, economic and cultural elements present in different localities" (Sterr et al., 2000, p. 5). Therefore, a conceptual framework for coastal vulnerability assessment was suggested to overcome the deficiency of the existing methodologies (Fig. 6.4). The framework is divided into the socio-economic and the natural system, since the assessment has to incorporate both sides of the coin. Climate change and the consequential sea level rise impose effects on both systems, and both systems have different means of reacting or adapting to changing circumstances. Sterr et al. (2000, p. 14) concluded that "coastal adaptation requires data and information on coastal characteristics and dynamics, patterns of human behavior, as well as an understanding of the potential consequences of climate change. It is also essential that there is a general awareness amongst the public and coastal planners and managers of these consequences and of the possible need to act".

Integrated Flood Risk Management

Coming back to the FloodSite project *flood risk management* has been defined as holistic and continuous analysis, assessment and reduction of flood risk consisting of different compartments: *risk analysis*, *risk assessment* and *risk reduction* – see Schanze (2006).

First step is to determine the hazard which could threat the area either by fluvial, pluvial or tidal flooding. Within the determination the probability should be the main aspect of investigation. In case of a failure of a dike, for example, the topography is a crucial parameter for the exposure to inundation. Based on the results of the hazard determination the vulnerability has to be assessed. In the previous section the process of vulnerability assessment has been touched. Finally, the combination of hazard and vulnerability determination will lead to the estimated risk.

The compartment of risk assessment comprises the perception and the weighting of risks. The process of weighting risks will not be touched in detail in this book, but elements and aspects discussed in Sect. 7.3 could be used to initiate such a process.

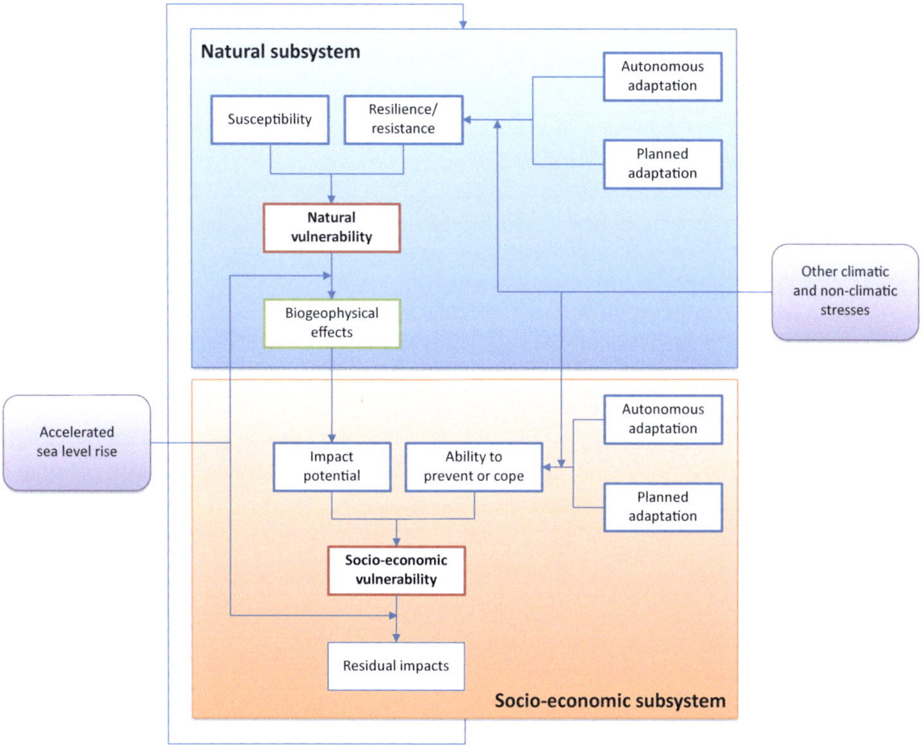

Fig. 6.4 Conceptual framework for coastal vulnerability assessment. Source: adapted from Sterr et al. (2000)

Risk Perception

"The day-to-day, "intuitive" perception and evaluation of risks (in short: risk perception) is a basis for acting and behaving in dangerous situations. It is also fundamental for decisions such as whether preventive protective measurements are taken or not" (Plapp, 2001, p. 2). Formerly, risk analysis and risk management based mainly on the technical or scientific approaches and were defined by the risk of technical constructions like nuclear power plants. The same is valid for coastal protection – for example, Giszas (2003); Kunz (2004), and RIKZ (2002) for The Netherlands.

> Scientists and decision-makers who assess and develop strategies to reduce the vulnerability of coastal zones to natural hazards and to improve disaster preparedness have to consider aspects in addition to quantitative measurable determinants such as inundation depth or the number of people affected. Whether a hazard has negative impacts on coastal societies or turns into a disaster depends to a great extent on human behavior. The human behavior in turn depends not on facts, but on perception, experience and knowledge. (Kaiser, 2006, p. 158)

Plapp (2003) carried out a comprehensive empirical investigation of risk perception of natural hazards (storm, high fluvial water, earthquakes) in different areas of Germany which revealed that the characteristics of the risk of these hazards are perceived as similar, for example, well known and of less personal vulnerability. Thus, the consequences a hazard can have or impose on the social system depends strongly on the personal reaction and preparedness to prevent damages; it is not only a matter of the political and administrative body.

> The risk perception is a fundamental base for the decisions and behavior concerning natural risks and their management of natural risks. Consequently, the risk perception of the inhabitants of a community has been taken into consideration concerning disaster management planning at community level. For the development of effective information strategies on protective measurements (risk and communications policies), the risk perception of the targeted group and as well influences on risk perception should be known. (Plapp, 2001, p. 2)

Taking these remarks into account, the results of the investigation conducted by Kaiser (2006) have shown that risk perception is a function of time, but not necessarily correlated with higher awareness resulting in better self-preparedness or precautionary action. People who experienced the disasters of 1953 or 1962 are more aware of the risk living in low-lying areas, but this did not necessarily lead to precautionary measures (Kaiser, 2006). A comprehensive study about risk perception and the detailed results for the southern North Sea can be found in Kaiser et al. (2004).

Recently, the multidisciplinary research project "Climate Change and preventive Risk and Coastal Protection Management at the German North Sea coast (KRIM)" studied the questions: What are the requirements of coastal protection to be integrated in ICZM under the boundary conditions of an accelerated sea level rise and increased storminess? And what are the interpreting pattern and decision-making processes that influence the integration process? The KRIM project was divided into three parts: (a) political-administrative risk, (b) scientific risk, (c) public risk (Schuchardt and Schirmer, 2007).

The main results of the sub project "Climate Change and Public" are congruent with the findings published by Kaiser (2006). Almost 95 % of the interviewees agreed that the consequences of climate change will force increased efforts in coastal protection (Peters and Heinrichs, 2007). "The analysis of cognitive reactions of test persons reading an article about coastal protection confirms the risk perception of storm surges as a relevant risk and a high confidence in present coastal protection" (Peters and Heinrichs, 2007, p. 127–128).

Another compartment comprises the reduction of risk, which can be applied prior to, during and after the flooding event. The intention of coastal protection or, in general, water management comprises the avoidance of disastrous impacts of flood events (pre-flood reduction). Contingency planning and emergency aid cover the part of the reduction of impacts of flood events, which means to reduce the consequences after the failure of safety infrastructure either by physical measures or by financial and communicative activities. With post-flood reduction in the last part of the compartment risk reduction the recovery after an event happened is meant. After completion of the immediate emergency aid the phase of recovery has to start which is called the post-flood reduction.

Coming back to risk classification after WBGU, an important question concerns the possibility of transferring one risk type (e.g., cyclops) from a certain risk area to another. Figure 6.5 shows that the transfer of risk types within a certain areas as well as from one area to another might be possible. Consequently, the question is: If storm surges are of the cyclops type, what has to be done to change the risk area? While the probability of occurrence of a storm surge cannot be influenced, it is necessary to reduce the extent of damage (see Fig. 6.5). This can be achieved by precautionary measures that enable the transfer to the normal risk area. But what are the consequences of this approach to coastal protection?

Risk according to coastal protection can be defined as introduced by formula (6.1). For the determination of the "probability of failure" the design water level and the construction of the embankment have to be considered, for example, Kortenhaus (2003). In case of failure the vulnerability is linked to the magnitude of the damage. The magnitude of

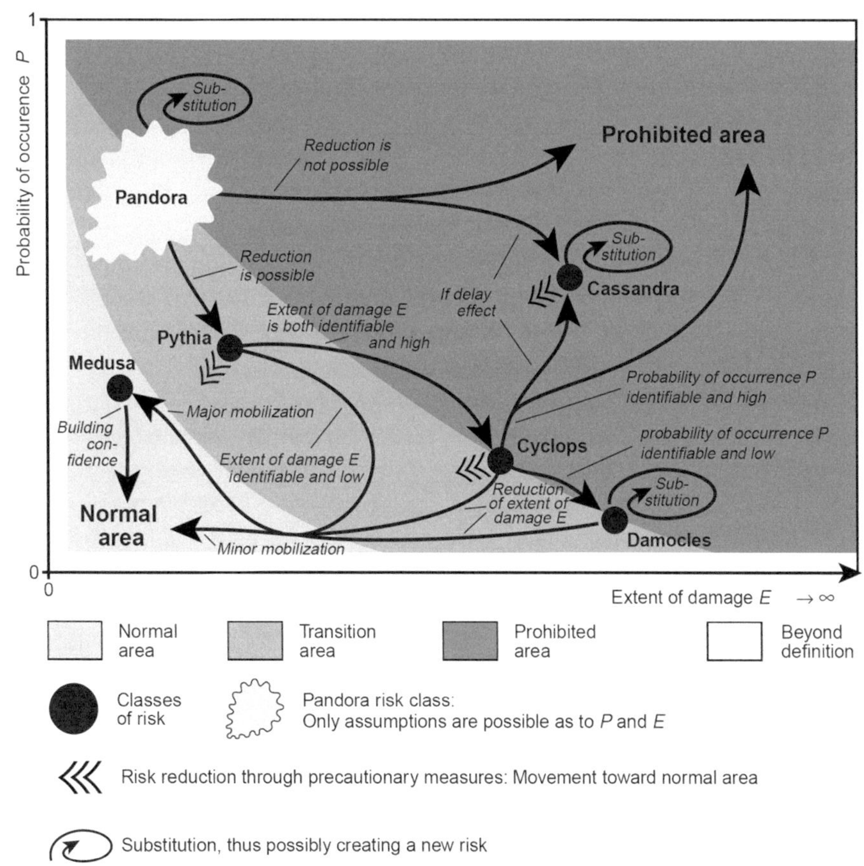

Fig. 6.5 Risk dynamics. Source: WBGU (1999)

the damage depends on several items some of which have already been mentioned. The design water level has to be calculated as described in Sect. 5.2. The significance of these uncertainties inherent in the determination for decisions on coastal protection are comprehensively described, for example, in Kunz (2004). Associated with the design water level is the probability of failure of a main dike; this probability should be as low as possible and it should be the same for the entire German coast.[3] The same probability of failure leads to different risks because there are differences in the vulnerability of rural and urban areas behind the dike. In the Netherlands the concept acknowledged the different degree of vulnerability with the different safety levels ranging from 1 to 10,000 (densely inhabited Delta area) to 1 to 1250 along the rivers (Fig. 5.6 on p. 71).

The design water level in Germany is the result of the societal decision-making process. The duty of coastal protection is the protection of people and their property against flooding. This societal demand implicates that the embankments should be built as safe as possible. But the question is how to define the term "as safe as possible"? Answering the question requires to consider that the flood-prone area has to be protected against flooding by a technical structure (i.e., dike) which height is determined by taking into account the highest storm tide, the wave run-up and the fact that guarantee via absolute safety is not possible. Obviously, the main problem is that the highest storm tide cannot be determined and all attempts to do so display remaining uncertainties. A safety standard has to be defined which is able to fulfill the subjective impression of safety for the public and provide an applicable safety standard for a construction. Consequently, the safety standard is not defined for the flood-prone area, it is defined for the technical structure. Over the last decades the safety standard, as defined by the Engineering Committee (Engineering Committee, 1962), has met public requirements. The safety standard is incorporated in a protection concept which guarantees a defined safety or an accepted risk and reasonable costs for the implementation (Kunz, 2004).

In this context, Giszas (2003) and Kunz (2004) refer to the ALARP concept (As Low As Reasonably Practicable), which is a modified version of the risk classification by the WBGU. The ALARP concept divides the risk into three main groups: (a) acceptable risk, (b) tolerable risk and (c) non-tolerable risk. The ALARP area is the area between acceptable risk on the one hand and non-tolerable risk on the other hand. Kunz (2004) divides the technical part of the protection approach into:

(a) definition of the critical (decisive) high water level,
(b) definition of the decisive wave run-up and acceptable over-topping rate,
(c) periodical inspection and improvement of safety status,
(d) technical standards and recommendations for the construction and maintenance of coastal protection infrastructure and
(e) reduction of risk by the introduction of additional embankments.

[3]Legally binding specification in the State Law on Dikes in Lower Saxony, Germany.

Non-technical measures have to be considered such as contingency planning and public information. At the moment, the concept of coastal protection at the German North Sea coast, that is, the design of a single line, leads to the strengthening and heightening of this line. This is caused by items (a) to (d) of the technical concept. Item (e) has been acknowledged by the State Law on Dikes (§29), but in practice this is not enforced (see Sect. 5.2). The expectations of the inhabitants and the politician in the flood-prone are that the agreed safety standard will be guaranteed, which under changing circumstances lead to the necessary adjustment and improvement of the embankments, that is, mainly heightening of dikes. If this reactive management strategy is not be enhanced by additional measurements in the hinterland, additional problems might occur (Kunz, 2004, p. 268):

- The reactive strategy will cause an increase of vulnerability, because of increasing high water levels and the failure of a main dike can not be eliminated. Further development in the flood-prone area will also increase the vulnerability.
- The concentration on the limited budget caused the heightening of the main dike and the disregard of the second dike lines.
- The strategy to continuously improve the existing line of defense to increasing sea levels will tie up almost all the available budget.
- The current strategy implicates the impression that additional measurements like self-preparedness in case of failure are unnecessary.
- The tight governmental budget will give rise to the argumentation that no alternative to the current strategy is available.
- According to international agreements the reorientation to integrative management needs to consider the protection of the people, the development and the conservation of values and the principle of sustainability. For coastal protection a shift of paradigms from "defense" (single line of defense) to risk management (spatial protection concepts) will be necessary. The consequence will be that the safety standard has to be adjusted to the required sustainable development of a certain area.

Although the risk approach has not been applied in Germany, for example, to the determination of the design water level yet, due to many uncertainties, these approaches are already the subject of basic scientific research – see, for example, Oumeraci and Kortenhaus (2002); Kortenhaus (2003); Elsner et al. (2004); Mai (2004) and Kortenhaus et al. (2007).

According to the current implementation of the FRMD an integrated risk management approach is required. Within the KRIM project an integrated risk management approach was applied to the Lower Weser region, that is, improving cooperation and collaboration between different agencies responsible for risk reduction in the coastal zone, especially for coastal protection. Markau (2003) suggests a risk concept to deal adequately with natural hazards in coastal regions: (a) scientific risk analysis (investigation of states), (b) socio-political risk evaluation (assessment of social effects) and c) economic-political risk management (optimization of risk situation) – similar to (Schanze, 2006). The main issue is that these different approaches do overlap in respect of information and applied methods. Cooperation and communication is limited because these fields stick to their traditional scope of responsibilities. The disadvantages of the sectoral approach are the divergent aims

Fig. 6.6 Brush wood groynes reducing wave impacts and, though enhancing sedimentation in order to enable mud flats to grow with sea level rise (subsequently reducing risk at flooding). ©Frank Ahlhorn

and the difficulty in building a consensus on certain tasks in planning. The advantages of the integrative risk approach are the early cooperation on and communication of problems and challenges. Also the reduction of the knowledge deficit and exchange of information to tackle the risk is rated as an advantage (Markau, 2003). Problems or obstacles within an integrated approach are the possible interruption or the intentional time delay of the process by an involved institution.

Markau (2003) suggests to divide the sectors into analysis, evaluation and management within the integrative approach because each sector has its specialists providing high quality data and information for the process. This approach has been applied at the KRIM project, and the experience was that the effort needed to generate cooperation between the different sectors should not be underestimated. Efficient and effective cooperation and communication between all parts requires a big effort.

A basic insight into the necessity of participation is given in Renn et al. (2005, p. 99):

> The larger the number of individuals and groups that have the opportunity to actively participate in risk regulation, the greater is the chance that they develop trust in the institutions of risk regulation and also assume responsibility themselves. However, participation cannot and must not substitute for effective and timely risk management. It should proceed parallel to and along with the prescribed regulation process. Above all, the participation process should not obscure or diminish the responsibility of legal decision makers. Participation within the framework of collectively binding regulations serves to prepare and help in decision-making processes, but not to distribute the responsibility among many (if possible anonymous) shoulders.

To conclude, the advantages and disadvantages of integrated approaches are valid for many projects dealing with cross-sectoral tasks. These experiences have been made and published within risk management, disaster management and also in the investigation of multi-functional coastal protection zones (Fig. 6.6) – see, for example, Ahlhorn (2009). Therefore, within the process of the implementation of ICZM projects the focus has to be on integration and in particular on cooperation and communication.

6.1.3 European Flood Risk Management Directive

Between 1998 and 2009 major damaging floods occurred in the European Union, including the catastrophic floods at the Rhine river in 1993 and 1995, or at the Elbe river in summer 2002. These floods caused approx. 1126 deaths, the displacement of about half a million people and at least 52 € billion in insured economic losses (EEA, 2010). Severe floods in 2005 further reinforced the need for joint action.

The Flood Risk Management Directive (FRMD) was proposed by the European Commission in early 2006 (EC, 2006) and was finally published in the Official Journal on 6 November 2007 (EC, 2007). Its aim is to reduce and manage the risks that floods pose to human health, the environment, cultural heritage and economic activities. The Directive forced Member States to carry out a preliminary assessment by 2011 to identify the river basins and associated coastal areas at risk of flooding. For such zones they needed to draw up flood risk and flood hazard maps by 2013 and establish flood risk management plans focused on prevention, protection and preparedness by the end of 2015. The FRMD applies to inland waters as well as all coastal waters across the whole territory of the EU.

The FRMD shall be carried out in coordination with the WFD, mutually by the coordination of preparing flood risk management plans and river basin management plans and through coordination of the mandatory public participation (see also 6.1.1). All assessments, maps and plans prepared have to be made available to the public.

Objective

The FRMD includes the following obligations for the Member States (Articles 4 to 10, see EC 2007):

Preliminary flood risk assessment: Action will only be taken in areas where potential significant flood risks exist or are reasonably predictable in the future.

Flood hazard and flood risk maps: Flood risks would be mapped for the river basins and sub-basins with significant potential risk of flooding, in order to

- increase public awareness;
- support the process of prioritizing, justifying and targeting investments and developing sustainable policies and strategies;
- support flood risk management plans, spatial planning and emergency plans

Flood risk management plans: Flood risk management plans are developed and implemented at river basin/sub-basin level to reduce and manage the flood risk. These plans include the analysis and assessment of flood risk, the definition of the level of protection, and identification and implementation of sustainable measures applying the principle of solidarity: not passing on problems to upstream or downstream regions and preferably contributing to reduction of flood risks in upstream and downstream regions.

To cope with the variety in flood events and impacts throughout Europe detailed objectives and deadlines for managing flood risks, for example, setting a Community-wide common level of protection which has to be achieved everywhere and within a certain time frame is not prescribed in the FRMD. This are at the responsibility of the Member States.

The appropriate level of protection may vary from river basin to river basin and even within each river basin (see Sect. 6.1.4). As flood risks may change over time due to climate change and changes in land use, it would be important to regularly review and, where necessary, update the three main elements of the FRMD.

Links to the Water Framework Directive

The FRMD includes a number of links to the WFD to ensure close coordination in the implementation processes. It is important to assure that there is no overlap of procedures and institutions and to generate synergies in the timetables for implementation.

For the implementation of the FRMD the Commission proposes that

> the administrative units shall be the same for the two Directives, i.e. the FRMD should be implemented on the level of the river basin districts (which includes not just river basins and sub-basins but also associated coastal areas) identified in the WFD Article 3 and the competent authority responsible for the WFD should also be responsible for the flood risk management actions. Furthermore, the implementation cycles and reporting mechanisms should be synchronized as regards the timetables, and it is envisaged that the Member States can choose to include the flood risk management plans in the river basin managements plans required under the WFD.

Implementation

Chapter II of the FRMD deals with the preliminary assessment of flood risk. In Article 4 a detailed description of the contents of the flood risk assessment process is provided. For example, it shall include a map of the river basin district including the borders of the river basis, sub-basins and where appropriate associated coastal zones showing topography and land use. A description of the flooding process and their sensitivity to change and a description of development plans is also required. Article 5 defines the categories the river basins, sub-basins or coastal zone should be assigned to: (a) for which no potential significant flood risk exist or are considered to be acceptably low or (b) for which it is concluded that potential significant flood risk exist or might reasonably be considered likely to occur. The assessment has to be done at latest 3 years after the date of entry into force, and the review shall be done at latest until 2018 and every 6 years thereafter.

Chapter III deals with the preparation of flood hazard and flood risk maps (Article 6–8). The categories covered in Article 6(3): (a) floods with high probability (once every 10 years), (b) floods with medium probability (once every hundred years) and (c) flood with low probability (extreme events). For these categories necessary information should be given as follows: projected water depth, the flow velocity, where appropriate, and the areas which could be subject to bank erosion and debris flow deposition. The third part of Article 6(5) deals with indicative flood risk maps which should contain: (a) the number of inhabitants potentially affected, (b) potential economic damage in the area and (c) potential damage to the environment. Article 6(8) formulates that the flood risk maps should be finished until 2013 at latest and should be updated every 6 years thereafter.

Chapter IV deals with the flood risk management plans, the third pillar of FRMD (Articles 7 and 8). These articles are dealing with the development and implementation of flood risk management plans in vulnerable river basins and coastal areas as well as the coordination mechanisms of the management plans within the river basin districts (EC, 2006). An important aspect is mentioned in Article 7(1): the Member States shall prepare and implement flood risk management plans at the level of the river basin district for the river basins, sub-basins and stretches of coastline identified under point Article 5(1). And in Article 7(2): Member States shall establish appropriate objectives for the management of flood risks for the areas identified [...], focusing on the reduction of potential adverse consequences of flooding and for human health, the environment and economic activity, and, [if considered appropriate, on non-structural initiatives and/or on the reduction of the likelihood of flooding.

6.1.4 The Dutch Flood Risk Management Approach

Referring to the description in Sect. 5.1.4 on the development of the Dutch approach to coastal protection the paradigm shift from probability of exceedance to probability of flooding will be elaborated. The probability of exceedance introduced after the storm surge in 1953 and manifested by determining safety standards for assigned dike-rings[4] according to their vulnerability against flooding led to a continuous construction program for heightening and strengthening of dikes (main dikes at the coast as well as levees at the rivers). The Flood Defense Act (see TAW, 1998) in 1996 confirmed these safety standards (see Fig. 5.6).

Due to fluvial flood events along the major rivers also a shift of the approach at rivers took place which led to the program "room for the river" (RWS, 2012, 2013). Besides the usual measures of heightening and strengthening of river dikes some more innovative measures were introduced (RWS, 2012, p. 6):

- *Floodplain excavation*: By removing layers of soil from certain parts of the floodplains, more room is created for the river when the water level rises,
- *Relocation of dikes*: By relocating dikes further away from the river, the floodplains become wider.

[4]Dike-ring area: An area that must be protected by a system of dikes against flooding, in particular high storm surges, high surface water in the major rivers or high water in the Ijsselmeer or a combination thereof (Flood Defense Act, Section 1).

- *Depoldering*: The dike on the riverside of a polder is relocated further away from the river. This depolders the area and enables water from the river to flood this area at times of high water levels.
- *Riverbed excavation*: The riverbed is made deeper by taking away the top layer of the riverbed. Because the riverbed is lower, there is more capacity for river water.
- *Water storage*: As a result of the controlled combination of a closed flood gate and large volumes of river water flowing to the sea, a specific retention polder serves as an area of temporary water storage.
- *High water channel*: A high water channel is formed between two dikes. The channel branches off from the river and transports some of the river water via another route at times of high water levels.
- *Lowering breakwater spurs*: Breakwater spurs ensure that the river does not alter its course or lose depth. However, at times of high water, spurs slow down the flow of the water. By lowering them, the water has a better chance of being conveyed faster.

These measures will be implemented as tailor-made measures in about 30 projects along the major rivers in The Netherlands.

The Floris research project (see Sect. 5.1.4) was launched to investigate the probability of flooding of the existing dike-rings in The Netherlands. This project developed and applied a risk-based approach to flood safety which should offer the prioritization of measures in flood protection. Until that time the Dutch strategy was mainly based on the probability of exceedance and the consequences of over topping (see Sect. 5.1.4). Several recommendations in the past already acknowledged that many more failure mechanisms exist. The task was to analyze and assess the extend of the probability per dike-ring consisting of multiple and different protective elements. According to the slogan *the chain is as strong as the weakest link* the dike-rings were further divided into dike sections.

In 2008 the Delta Commission published new recommendations for water management in The Netherlands which again recommended to apply a risk-based approach (Delta Commission, 2008). The next step was the implementation of the Water Act in 2009 where the safety standards per dike-ring were manifested again.[5] In §2 section 2.2(1) it is specified that "[...] for each dike ring as the average annual overtopping probability of the highest high-water level that the primary flood defense structure erected as direct defense against external waters, must be designed to withstand, also taking in to account *other* factors that determine the flood defense capacity of such structures." That means that in the Water Act a shift slightly took place not to only rely on the probability of exceedance. "The safety standard for each dike ring shall be further specified by ministerial order as the average annual probability of flooding of the area protected by the dike ring as a result of the breach of a primary flood defense structure" (§2 section 2.2(2)).

[5]Here defined as: system of primary flood defense structures that, either alone or in combination with high ground, provides protection against flooding, in particular by external water (§1, Section 1.1).

In 2009 the National Water Plan was initiated and aimed at "The Netherlands, a safe and livable delta, now and in the future" (RWS, 2009, p. 4). The National Water Plan contained a 6 year program to achieve this main goal. In December 2014 the new National Water Plan for the period 2016–2021 was drafted containing the achievements so far and the ongoing measures. Furthermore, it contained the responsibilities in the water sector based on the Administrative Agreement on Water (2011) and the envisaged adoption of new standards for safety. As laid down in the draft (MinInfraM and MinEZ, 2014b, p. 10)

> the [Dutch] Cabinet aims to achieve the following goals:
>
> - The flood risk management policy offers everyone living behind a dike in The Netherlands a tolerable risk level of at least 1 in a 100,000 per year. This means that the probability of dying as a result of a flood for any individual should be no greater than 0.001% per year.
> - Moreover, additional protection is offered in areas where there may be:
>   - potentially large groups of victims;
>   - and/or major economic damage;
>   - and/or serious damage as a result of the failure of vital and vulnerable infrastructure of national importance.

Consequently, another change took place: The new standards are no longer valid for dike-rings they are valid for dike-sections[6] – see also Eijgenraam (2006, 2007). According to the new classification it was divided into six categories ranging from 1/300 per year to 1/100,000 per year. The Netherlands aimed at the conduction of "anticipatory assessment and ensure a robust design" (MinInfraM and MinEZ, 2014b, p. 11) of protection against flooding.

The basic idea is to introduce the Multi-layer Safety (MLS) concept which comprises three layers:

Layer 1 – Preventive measures to limit the probability of a flood.

Layer 2 – Spatial organization of an area to limit the consequences of a flood and, in specific cases, to contribute directly to the desired level of safety.

Layer 3 – Disaster management to limit the consequences of a flood in terms of casualties.

The MLS is aiming at acknowledging the full range of (coastal) flood risk management from prevention to recovery based on the risk definition provided by the formula (6.1) on p. 130 (see also Fig. 6.7). The recently published report for the new Delta program

[6]In the Floris project the following definition was applied (see Vergouwe, 2015): A levee system consists of all kind of flood defenses. In the Floris project each system was first divided into homogeneous stretches of some 250 to 1500 m. Hydraulic structures like locks and pumping stations were treated as separate parts.

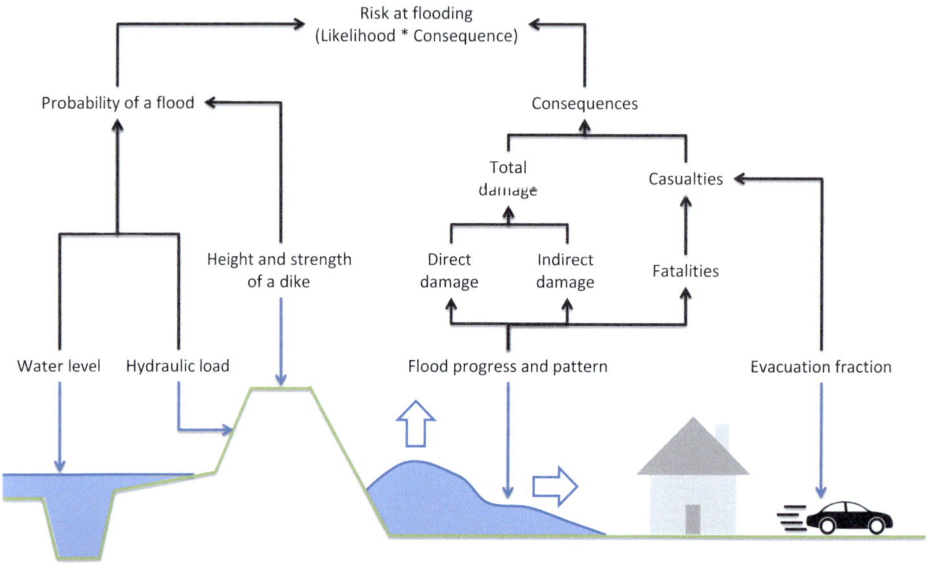

Fig. 6.7 Risk-based approach which will be applied in the future in The Netherlands. Source: adapted from MinInfraM and MinEZ (2013)

elaborate on these issues – see MinInfraM and MinEZ (2013, 2014a, 2015); Vergouwe (2015).

6.1.5 Conclusions

This chapter explained the transition from a traditional (mainly sectoral) approach of drainage and flood protection management towards a more integrated water management comprising the qualitative (i.e., concerning the WFD) as well as the quantitative (i.e., concerning flood risk management and the FRMD) aspect. With the introduction of the European Water Framework Directive (WFD) a more integrated approach was applied to improve the ecological and chemical status of Community water bodies. The WFD is comprising and substituting different older European and national policies and made also recommendations on the implementation of the goals, for example, by guiding principles for stakeholder participation in the process of identification and application of measures. The WFD does not explicitly touch the aspect of quantity of surface waters, so the several years later the Flood Risk Management Directive (FRMD) was published. Within the FRMD a risk-based approach for the protection against high water levels (either at rivers or at the sea) is proposed by introducing flood risk assessment, the preparation of flood hazard and flood risk maps and, finally, by demanding to prepare flood risk management plans. Both Directives follow a consecutive time schedule for the preparation of each step and the monitoring procedures. These Directives are called Framework Directives because they offer a framework implemented and enforced by each Member State. According to existing structures and legislation in Member States the way on how to implement and fulfill the requirements differ reasonably. Some Member States already have comprehensive administrative structures in place taking over the responsibility for river basins and their management, some are already running thorough monitoring systems on ecological and chemical parameters. If Member States are not in that situation they should have the chance to develop necessary structures and methods adapted and tailor-made for their context.

Regarding the FRMD different approaches exist on how to protect people against flooding. Some of them have been described in Chap. 5, for example, the Dutch and the German. The result of these different approaches seems to be rather similar, that is, high and strong dikes as protection against storm surges. The way how to determine the height or the structure of these construction differs significantly. The Dutch approach is based on the probability and is shifted in recent years to a more risk-based approach. The German approach, especially in Lower Saxony, is based on accomplishing safety not taking a risk approach. Although many parts of the chain for flood risk management do exist. These approaches have to be adapted according to the requirements of the FRMD. In Germany, flood hazard or flood risk maps are existing, but the information content is disappointing, because the currently published maps are calculated by subtracting the elevation of the area from the respective design water level of this area. The result should indicate the height of the water level in that area in case of flooding (e.g., dike failure). Neither the topography nor existing infrastructure was taken into account, because fundamental data is partly missing or the available methods do not fulfill legal requirements in German. Until now the flood prone area is determined as the area which is protected by the main dikes. It is based on the calculated design water level of the respective area which is then connected to a certain contour line (e.g., if the design water level is approx. 5 m so the inland border of the protected area is confined by the 5 m contour line).

Subsequently, these different approaches have to be adjusted to the requirements of the FRMD. Following the Dutch case it is easier, as shown in this chapter by enhancing the probability-based approach to a risk-based approach (see Sect. 6.1.4). Although, for the full transition to a risk-based approach many challenges have to be tackled.

Exercise

1. The description of the WFD on p. 121 provide a first overview on the development background and the intention of each Directive. Find information on the current status of implementation of both Directive in selected Member States.
2. Compare the approaches of your selection. What are the differences, what are the similarities? Which Member States still have not implemented the FRMD?
3. The FloodSite project has briefly been touched to introduce the term flood risk management on p. 132. Flood risk management consists of three compartments

Coastal Adaptation Strategies	Adaptation Options	Adaptation Responses	Examples
Protect	Increased robustness	Advance the line	Land reclamation
		Hold the line	Beach nourishment
Accommodate	Increased flexibility		Flood proof buildings, floating agricultural systems
Retreat	Enhanced adaptability	Retreat the line	Managed re-alignment
		Limited interventions	Ad-hoc sea wall
		No intervention	Monitoring only
	Reversing maladaptive trends	Sustainable adaptation	Wetland restoration
	Improved awareness and preparedness	Community-focused adaptation	Flood hazard mapping, flood warnings

Fig. 6.8 The first proposed IPCC adaptation strategies were linked to certain adaptation objectives. These objectives were split up in various adaptation responses. The examples indicate the diversity of possible measures to be developed for the adaptation to sea level rise. Source: adapted from Parry et al. (2007)

according to the definition of the FloodSite project. Compare the flood risk management approach of the FloodSite project with the safety approach of Lower Saxony (Germany) described on p. 73.

4. The three strategies proposed by IPCC for the adaptation to sea level rise are retreat, accommodate and protect have been enhanced in the 2007 IPCC report – see Fig. 6.8 and Parry et al. (2007). Are there more adaptation objectives and responses conceivable? Find more examples for each strategy.

6.2 Integrated Estuary Management

6.2.1 Introduction

The intention is to provide examples to show that in many estuaries new path are struck according to integrate different interests and needs rather than developing an integrated estuary management concept. Looking back at Sect. 5.4 where the historical development and the prioritization of the easiness and safety of shipping in estuaries was described, here, examples are given which will work on the experienced deficiencies. Straightening and narrowing of river channels affected the hydrodynamic and morphological system of the estuary. Parts are suffering siltation while others are eroding. River conservancy works were applied to enhance and improve the properties of the river for shipping. With the emerging interests and needs of environmental and nature conservation a shift in thinking about estuaries evolved. More and more also environmental aspects have to be considered in estuarine management, for example, in Europe because of the introduction of the European Water Framework Directive and, of course, prior to that over decades based on the development in Member States.

The main issues in estuaries are the tidal amplification together with morphological changes and salt water intrusion, as it has been identified in the EMOVE project – see also Gregory (2006); Ducrotoy (2010) and EMOVE Project Consortium (2015a). Therefore, selected approaches will be provided on how estuarine managers or administrative bodies are trying to deal with these challenges. The ideas to counteract tidal amplification of an estuary will be exemplified for the Elbe river. Tidal range doubled (tidal gauge St. Pauli) since human interference in the river Elbe started, and the tidal characteristic is disturbed. The flood current is stronger than the ebb current resulting in a net import of sediment into the estuary. Dredging is essential in estuaries to keep the fairway to the desired draft. Thus, the dredged sediment has to be handled either by extracting it from the river system or relocating it within the system. Different approaches exist depending on the characteristic and the shape of the estuary. Here, the example of the Schelde river will illustrate that traditional sediment management, that is, dredging and disposal of the material outside the system, might show undesirable effects. Deepening and straightening of river channels led to an increased flow velocity, hence, to an upstream shift of the brackish water zone. Different types of land use adjacent to the estuary may depend on fresh water abstraction from the river. If the brackish water zone is moving upstream abstraction of fresh water might be limited or impossible. The situation of salt water intrusion, the consequences and possible solution options will be exemplified for the river Weser in Germany.

6.2.2 Ideas to Influence Tidal Amplification

Over centuries the way of adapting rivers to an increasing demand by transport was deepening and straightening of river channels (see Sect. 5.4.3 and Fig. 5.22). As major consequence tidal range increased up to twice as high as it was under natural conditions. Furthermore, the normally ebb-dominated turned to flood-dominated estuaries which means that sediment is imported into the upper reaches of the river. For example, the harbor of Hamburg suffers sedimentation of fine grained material in the harbor area (Fig. 5.33) which lead to the reduction of the possible draft and worsened the turbidity of the water column. Instantaneous measures were and still are high efforts in maintenance dredging and the implementation of artificial river conservancy measures. The closure of former tributaries to the Elbe river and side channels aimed at concentrating the current in the fairway and to increase the flow velocity for an aspired self-sustaining effect. Finally, the current cross-section of a river looks like shown in Fig. 5.22 (bottom). Continuously high efforts and to some extend an even increasing demand in dredging result in developing new ideas and to some extend to scrutinize the existing measures.

Today, the way of dealing with estuaries is to almost return to the picture like in Fig. 5.22 (top) shown. The width of the cross section should be enhanced rather than the draft should be reduced, because container vessels show a constantly increasing size. Dike relocation where possible and feasible and the re-connection of former side channels are

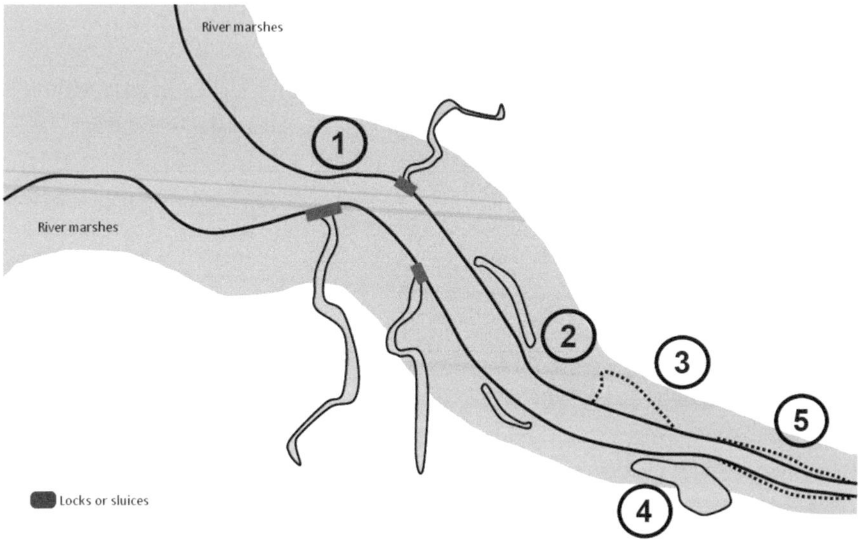

Fig. 6.9 Ideas for the improvement of hydrodynamic and morphological aspects of an estuary: ①: Re-connection of tributaries, ②: Re-connection of old side channels, ③: Relocation of dikes, ④: Connecting standing water bodies, ⑤: Widening of the river channel. Source: adapted from Freitag et al. (2007)

ideas for the future to improve hydrodynamic and morphological aspects of estuaries. To reduce the vulnerability against flooding an extension of the tidal capacity is envisaged by widening the cross section or by re-connecting former tributaries or at least manage the locks and sluices differently (Fig. 6.9).

The implementation (or better: return) of these measures might not be easy because the necessary area behind the river dikes is developed and different types of land use are in place. For example, industry, agriculture infrastructure and settlement were erected. At some places densely populated areas exist and at other places highly vulnerable infrastructure (e.g., nuclear power plants) was installed. But, nevertheless human interference show considerable impact in the hydrodynamic and morphological development of an estuary and according to existing environmental legislation counteracting against some changes is demanded – see, for example, Gregory (2006); Ducrotoy (2010) and TIDE Project Consortium (2013).

6.2.3 Example: Morphological Management

In Sect. 5.4.4 the changes of the multiple channel system of the Schelde were described (Fig. 5.30). Over the last decades the multiple channel system turned mainly into a two channel system. One channel conveys the flood current and the second by the ebb current (Fig. 6.10). Due to differences in flow velocity and other circumstances structure and shape

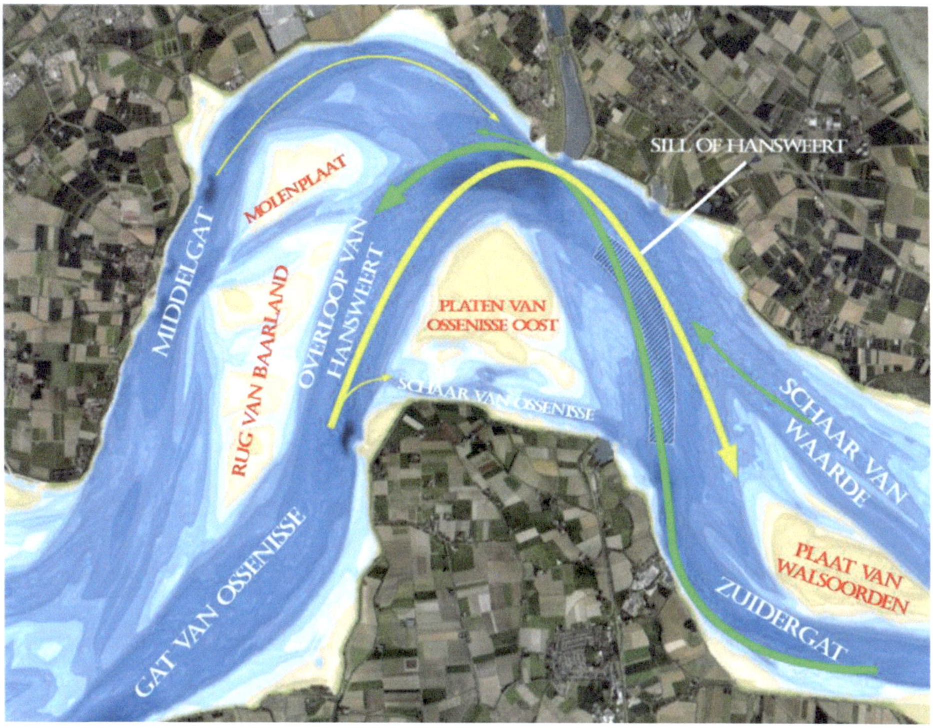

Fig. 6.10 The multichannel system of the Schelde consists of ebb- and flood-dominated channels. The *yellow arrow* shows the flow-direction of the flood current and the *green arrow* shows the flow direction of the ebb current. Sand plates intersect the channel system which suffer erosion. Important estuarine habitats are either drowning or diminish. Eroded material is disposed within the channels. Source: APA (2012)

of the channels vary and flood-dominated channels are more shallow. A detailed description of the morphological development at the Schelde and the prospects would go beyond the scope of this book, but can be found, for example, in Mol (1995); Wang et al. (2002); Jeuken et al. (2013); Coen (2008); Jeuken and Wang (2010); APA (2012); Deltares (2013); Vandenbruwaene et al. (2013); Winterwerp (2013); VNSC (2013, 2014) and IAHR (2015).

Two main transport directions of sand exist in the estuary: From the outer reach (near the mouth) sand is exported into the North Sea. In the upper reaches sand is transported upstream to the harbor of Antwerp. For silt the situation is different, it is imported through the whole length of the estuary and partly accumulates at the harbor of Antwerp (VNSC, 2013).

To effectively manage the morphological development of the Schelde so-called *morphological cells* were defined (Fig. 6.11, e.g., Jeuken and Wang 2010; Deltares 2012). An estuary consisting of a multiple channel system is in the transition between completely drowning without inter tidal areas or the opposite complete siltation, depending on specific

Fig. 6.11 The Schelde is intersected by sand plates and consists of multichannel system. Sediment is eroding and mainly relocated in so-called morphological cells. In the Schelde seven morphological cells have been identified. Within these cells the approach of morphological management will be applied by relocating the dredged material within these cells and not extracting from the system. Source: Deltares (2012)

circumstances such as sea level rise and sediment budget. Although in a transition stage for decades this stage might represent a dynamic equilibrium. Subsequently, the Schelde estuary consists of a chain of morphological cells. The sediment budget of these cells differs and mainly depends on the gross sediment transport (Wang et al., 2002; Winterwerp, 2013).

On the basis of these findings a morphological management concept was proposed (e.g., Deltares, 2012; Winterwerp, 2013). The dredged sediment from the fairway to keep the necessary draft is relocated in the respective morphological cell. The material is not extracted from the system it is used to counteract undesired negative effects of eroding inter tidal or shallow water areas. Sand plates are eroding and the material is deposited either in the main or in the secondary channel. Thus, important ecological habitats are lost and the channel, especially the fairway is silting up.

Different measures are proposed to counteract the undesired morphological development. These measures are divided into *soft* and *hard* measures. *Soft* measures consist of the flexible disposal of material in very deep parts of the main channel or on specific inter tidal areas. New channels might be dredged to connect the ebb and the flood channel, or specific inter tidal areas will be lowered to reduce tidal range. *Hard* measures comprise the construction or the removal of hard infrastructure such as embankments and groynes (EMOVE Project Consortium, 2015b).

6.2.4 Example: Solution Approaches for Salt Water Intrusion

Introduction

An introductory description of the hydrological and geomorphological situation of the Weser estuary was provided in Sect. 5.4. According to the Water Framework Directive an Integrated Management Plan has been developed for the Weser – see FGG Weser (2009).

To exemplify the problem of salt water intrusion in estuaries the northern part of the county Wesermarsch is selected as example, that is, the municipalities Butjadingen (approx. 129 km^2) and Stadland (approx. 113 km^2). The region comprises two types of land-marine transition zones (i) the Weser estuary and (ii) the Wadden Sea with the unique Jade Bay (see Fig. 6.12). Surrounded by the North Sea and the river Weser, especially in the northern part, and with a coastline of 160 km the Wesermarsch is characterized by an almost prevailing saline environment. Also the ground water body is strongly influenced by salt water penetrating underneath the peninsula. (Fig. 5.40). As shown in Sect. 4.2.2 and Fig. 4.6 the transition zone between saline and fresh ground water is oscillating.

Two-thirds of the whole county area (822 km^2) are situated below mean high water level, which requires coastal protection by dikes and a comprehensive drainage system consisting of ditches, canals, tidal gates and pumping stations. Nearly 80 % of the Wesermarsch is characterized by agricultural use, besides tourism and harbor industry it is one of the most important economic sectors in this region. Marsh and peat soils occur predominantly in the county, 90 % of the agricultural land is under grassland farming (meat and milk production).

Problem Description

Over centuries water management in the region Wesermarsch was affected by alternating flooding events and land reclamation, shaping the current coastline. Todays coastline has been established by building dikes over the last centuries while flood protection by dwelling mounds became less necessary simultaneously (e.g., Kramer, 1992). Dikes function as an artificial barrier between land and sea, which makes a natural water exchange impossible. Accordingly, the Wesermarsch region is drained during winter time to discharge the surplus of rainwater and watered in the late summer months to water the fields and cattle by ditches.

Water (drainage) boards are responsible for the maintenance of this water management system in the Wesermarsch. The drainage boards Stadland, Brake and Butjadingen are three of them (Fig. 6.12). The area of the drainage board Stadland is drained by ditches via the Strohauser canal and discharges the water directly into the river Weser. Pumping stations in the area of the drainage board are necessary to drain low lying areas. In Butjadingen six tidal gates are used to drain the area. The water is discharged into the North Sea, the Jade Bay and the river Weser (red circles in Fig. 6.12). Depending on the water level in the river and in the North Sea, the water has to be pumped or can be drained by gravitation. The watering of the agricultural land in both regions is performed by extracting fresh water from the river Weser.

Over the last decades problems regarding the drainage capacity in winter time and the quality of fresh water in late summer occurred – see, for example, Ahlhorn et al. (2010); Umlauf et al. (2011) and Bormann et al. (2012). More often the drainage system was not sufficiently able to store and discharge the increasing amount of rainwater. In summer time the salt concentration in the Weser exceeds the threshold for watering[7] due to an upstream

[7]The watering is mainly used to flush the ditches to prevent standing water bodies and, thus, harmful diseases, but is also used to feed the cattle. Therefore, salinity should not exceed a certain concentration.

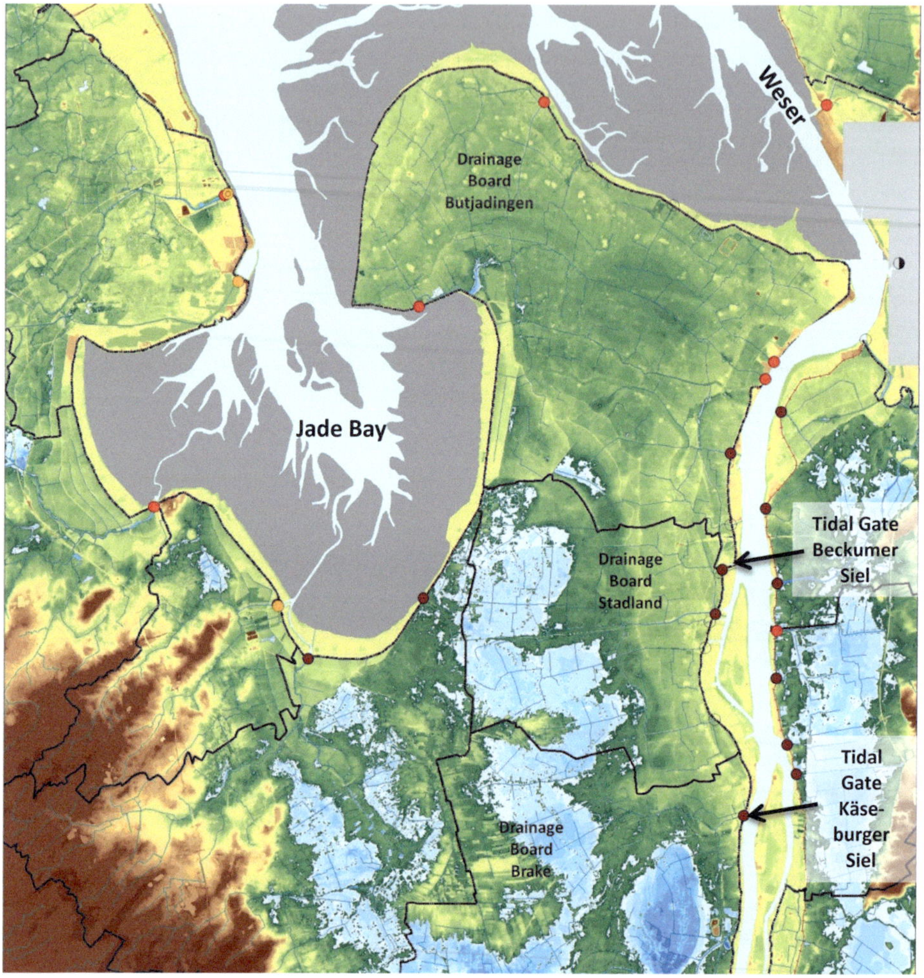

Fig. 6.12 North of the county Wesermarsch in north-western part of Lower Saxony (Germany). Three drainage boards of the county are directly linked to the river Weser. The northern boards (Butjadingen and Stadland) need fresh water from the river to flush the ditches in their drainage area. The tidal gate Beckumer Siel is the most southern point where fresh water is abstracted. Due to the deepening and straightening of the fairway the brackish water zone is almost oscillating around the tidal gate Beckumer Siel (*blue*: below sea level, *green*: 0–2 m above sea level, *yellow*: 2–5 m above sea level, *brown*: $\geq$5 m above sea level). Source: adapted from Ahlhorn et al. (2010)

movement of the brackish water zone, especially caused by deepening of the Weser and an increasing sea level (Kunz, 1995; Lange et al., 2008).

In future time, it is assumed that water management in the Wesermarsch will have to cope with an intensification of the present problems (Bormann et al., 2012). Climate change simulations for the Wesermarsch region projected an increase of water scarcity in summer, an increase of runoff formation in winter, and a rising sea level. Consequences may presumably be a longer watering period in summer and higher drainage demand in winter with concurrently less discharge ability due to sea level rise.

The deepening of the river Weser (Fig. 6.13), accompanied by a further upstream salt incursion of the river water, might reduce the availability of fresh water. Consequently, the watering and drainage system has to be adjusted and alternative solutions to solve these problems have to be discussed. Different sectors will be influenced by changes in water management such as agriculture, nature conservation and tourism.

Solution Options

Based on the results of the former EU Interreg IVB project Climate Proof Areas (CPA) solution ideas were jointly developed by the regional forum Wesermarsch[8] (Ahlhorn et al., 2011). Based on these results an investigation was started to enhance the list of possible solution ideas for the watering problem in the county of Wesermarsch. In the context of the EMOVE project this investigation has been conducted by applying the DPSIR approach. According to the DPSIR approach the identified solution ideas were named responses. The outcome of the investigation within the EMOVE project and the entire list of responses can be found in EMOVE Project (2015).

The main focus for the drainage boards is to improve and ensure watering and drainage is on technical solutions. For example, the extension of the Butjadinger canal to the south of the tidal gate Beckumer Siel, for example, down to the tidal gate Käseburger Siel (Fig. 6.12), hence, enlarging the existing watercourses. The water could be stored, distributed, and transported over a longer distance and could be available for a wider area for watering purposes.

This measure would affect different drainage boards, which supports the idea to reorganize the watering and drainage system across the northern drainage boards and to have trans-boundary cooperation between these boards. In the meantime, a joint venture has been established between the three drainage boards of Butjadingen, Stadland and Brake called Planungsverband Wesermarsch (Planning Association Wesermarsch).

To guarantee the livestock drinking water supply in summer time the existing drinking water system could be adjusted to the increasing demand. This means using high quality water to feed the cattle. A combination with water storage in winter time in reservoirs, for example, constructed close to the Beckumer Siel, might be a possible solution for having enough water in summer.

Associated with the currently planned fairway deepening of the Weser is an avoidance solution for the northern drainage boards in the Wesermarsch. This avoidance solution comprises an improved steering of the watering features and the heightening of the dikes around the Butjadingen canal which is feeding the fresh water into the drainage board areas up north.

---

[8]The regional forum was a stakeholder forum consisting of various representatives from different sectors such as agriculture, water management, spatial planning and nature conservation.

The adjustment of the water management system in the Wesermarsch region to climate-induced changes should aim at an effective and efficient watering and drainage system. To preserve the characteristics of the landscape as formulated in the development targets for the Wesermarsch region, the development of these possibilities have to be intimately connected with the future use of the rural landscape and should not have a negative influence on the environment. Finally, the concentration is set on two options:

- conveying fresh water from the Weser more upstream in the southern part of the county of Wesermarsch and
- implementing the avoidance solution after the final approval of the currently planned fairway deepening of the Weser

As the second solution option is going to be implemented when the fairway deepening would have been approved, the preliminary most promising option for the watering problem seems to the so-called *Master Plan Wesermarsch* developed by the respective drainage boards in the Northern part of the Wesermarsch.

The original idea of the Master Plan Wesermarsch was to improve the watering and the drainage system of the northern part of the Wesermarsch. The envisaged improvement should be benefiting for two major types of land use water management and agriculture. The agriculture sector is the main driver for the improvement of the watering system. The implementation of the Master Plan Wesermarsch should provide different positive effects for them:

- getting fresh water into the agricultural used areas of the drainage boards of Butjadingen and Stadland (and to some extent for Brake),
- enabling a continuously flushing of the ditches and canals to avoid standing water bodies (deterioration of the water quality by, e.g., private sewage features),
- maintenance of the existing ecological status of the ditches and canals with fresh (respectively brackish) water species for both flora and fauna,
- preservation of the existing NATURA 2000 sites in the county of Wesermarsch,
- preservation of the existing cultural landscape shaped by men in the last centuries.

The implementation of the Master Plan Wesermarsch will entirely contribute to the mission of the drainage boards. Nevertheless, it is mainly concentrated on the watering problem which is a special task of these drainage boards.

The current version of the Master Plan Wesermarsch has mainly been developed by representatives of the agricultural sector, but the original idea is much older. The water boards adopted this solution option, because, in the first draft, it provides benefits for the drainage tasks of these boards. Politics took the idea of the Master Plan Wesermarsch over and published information on funding issues via public press.

The main threats for the implementation are as follows:

- On political level the avoidance solution and the Master Plan Wesermarsch are closely connected, although these options have been separately developed. The avoidance solution is linked to the current planned fairway deepening and, therefore, has after the approval a legal status. The current planned fairway deepening of the Weser is at the court and politics might use the situation to postpone or delay the implementation process until a judgment has been rendered.
- The established Planning Association Wesermarsch and the representatives of the agricultural sector might fear that the active attendance in a participation process hamper or even freeze the funding of the Master Plan process.
- Main funding of the Master Plan Wesermarsch will be spread over three different levels and institutions, that is, Federal Waterway Administration and the Federal States Lower Saxony and Bremen. The Federal Water and Shipping Administration is responsible for the implementation of the fairway deepening in the Weser. The Federal States Lower Saxony and Bremen applied for the deepening to improve and maintain the accessibility of harbors.

The following improvements might help to get a step further within the implementation process:

- develop an integrated long-term vision,
- broaden the scope of the current project, try to incorporate other sectors and types of land use,
- assess future boundary conditions, to be prepared for future developments so that the proposed solution has long-term perspective,
- develop an alternative solution together with the involved parties from out the area, for example, enabling multiple purposes or/and get support by other parties,
- initiate a participatory integrated planning process to incorporate and merge the previous aspects,
- try to build new coalitions.

The benefits of the previously mentioned improvements might be as follows:

- If the Master Plan Wesermarsch is successfully implemented the Federal Water and Shipping Administration won't get further actions at the court for the next fairway deepening.
- A successful implementation of the Master Plan Wesermarsch might lead to the prospect that farming can be as it was. Consequently, the local inhabitants might be pleased.

- If future developments, especially regarding climate change, will be taken into account for the implementation, the region might be well adapted.

The Planning Association Wesermarsch and the Farmers Association should take actions to overcome the current deadlock. Several activities were conducted so far, a technical and economic feasibility study and a laser scan survey of the area of concern. Recently, hydraulic calculations were made to determine the necessary technical features for the water course, for example, to be able to bring the water to the northern part of the county Wesermarsch.

Although, there is a supporting political statement for the implementation of the Master Plan Wesermarsch some actions still have to be taken (the following items are neither comprehensive nor obligatory recommendations, they should be understood as compilation of ideas developed in a discussion process of the EMOVE project):

- If the deadlock is lasting longer, than the Planning Association Wesermarsch should think of trying to convince the local politicians, as representatives of the county Wesermarsch in the Federal State parliament.
- When the representatives of the Farmers Association as well as the Planning Association Wesermarsch are observing that the political support for the Master Plan Wesermarsch is not progressing, than they should consider to increase the public perception.
- Furthermore, it could be important to broaden the scope of the Master Plan Wesermarsch because focusing on engineering knowledge might be leading to shortcomings in other sectors or disciplines, for example, taking into account the requirements of nature conservation or tourism development.
- By broadening the scope of the Master Plan Wesermarsch new coalitions could be built and a wider basis could be founded.
- Try to convince the head of the county administration.
- Try to convince the state secretary of the ministry of environment.

On the other side of the river Weser also a shortage on fresh water is anticipated in the future. Due to the same reasons as for the drainage boards in the county Wesermarsch abstraction of fresh water from out the river Weser might not be sufficient (in quality) for the demand of watering. The tributary Lune discharges enough fresh water from the higher moraine area which could be pumped into the southern part of area. New bypasses between the river Lune and the existing water bodies could be constructed to feed the respective area with enough fresh water.

Almost the same situation, a further upstream oscillation of the brackish water zone, is expected at the river Elbe. In the low-lying marsh areas many fruit grower are located which need fresh water. Here, the idea is that dependent on the salt concentration in the river reach compensation will be paid if the impact of the encroaching saline water body forces the fruit grower to use drinking water for irrigation.

Fig. 6.13 Cruise liner at the quay of Bremerhaven. Deep going ships and vessels demand for sufficient water depth in the Weser estuary. Quay of Bremerhaven. ©Frank Ahlhorn

6.2.5 Conclusions

First approaches were introduced to walk along the path of integrating different interests and needs. For example with the morphological management to mutually improve the navigation in the main channel and the ecological status, for example, of inter tidal habitats in the secondary channels. Although, the priority for shipping and navigation lies on a desired depth for the fairway ideas emerged on how to improve the characteristics of the estuary by reconnecting former tributaries or former side channels of the river. Furthermore, to improve flood protection along the river more "room for the river" should be generated to increase the flood plain and, finally, lower water levels.

Exercise

1. Tidal amplification is a major effect of human interference in estuaries. Find information what tidal amplification really means for the estuary and how it is represented in different estuaries.
2. The concept of morphological management has briefly been described. What are the basic preconditions to develop a morphological management concept? What are the necessary and sufficient information needs?
3. Salt water intrusion is not solely a problem for estuaries around the North Sea, but worldwide. Find other examples and reasons for salt water intrusion in estuaries. What about salt water intrusion in the coastal aquifer? Does it also cause impacts on specific types of land use?
4. Only a small piece dealing with the issue of integration according to estuaries has been touched, for example, the part of how morphological or hydrodynamic measures influence the ecological and chemical status of a river has not been described. Take the report of TIDE Project Consortium (2013) as a staring point to find out more about various facets and needs in integrated estuarine management.

References

F. Ahlhorn. *Long-term Perspective in Coastal Zone Development. Multifunctional Coastal Protection Zones.* Springer, 2009.

F. Ahlhorn, J. Meyerdirks, and I. Umlauf. Speichern statt pumpen. Abschlussbericht. Technical report, University of Oldenburg, 2010.

Ahlhorn et al. Klimasichere Region Wesermarsch. Die Zukunft der Wasserwirtschaft. Technical report, University of Oldenburg, 2011.

[Antwerp Port Authority] APA. Morphological Management of Estuaries. Case Study of the Scheldt Estuary. Technical report, Antwerp Port Authority (APA), 2012. URL www.tide-toolbox.eu.

H. Bormann, F. Ahlhorn, and T. Klenke. Adaptation of water management to regional climate change in a coastal region – hydrological change vs. community perception and strategies. *Journal of Hydrology*, 454-455:64–75, 2012.

I. Coen. De eeuwige Schelde? Ontstaan en ontwikkeling van de Schelde. Technical Report D/2007/3241/203, waterbouwkundig laboratorium, 2008.

Delta Commission. Working together with water. A living land builds for its future. Findings of the Delta Commission 2008. Technical report, Deltaciommissie, 2008.

Deltares. Sand Balance Approach. Assessing sediment budgets and transports using bathymetric data. Technical report, TIDE Project Consortium, Delft, 2012.

Deltares. Water, zand en slib in het Schelde-estuarium. Technical report, Vlaamse-Nederland Schelde Commissie (VNSC), Delft, 2013.

J.-P. Ducrotoy. Ecological restoration of tidal estuaries in North Western Europe: an adaptive strategy to multi-scale changes. *Plankton Benthos Research*, 5:174–184, 2010.

[European Commission] EC. Directive 2000/60/EC of the European Parliament and of the Council of 23 October 2000 establishing a Framework for Community action in the field of Water Policy. Water Framework Directive, European Parliament and Council, 2000.

[European Commission] EC. Proposal for a Directive of the European Parliament and of the Council on the assessment and management of floods. COM(2006) 15 final, European Parliament and Council, 2006.

[European Commission] EC. Directive 2007/60/EC of the European Parliament and of the Council of 23 October 2007 on the Assessment and Management of Flood Risks. *Official Journal of the European Communities*, C288/27:27–34, 2007.

[European Environment Agency] EEA. Mapping the impacts of natural hazards and technological accidents in Europe. An overview of the last decade. EEA Technical Report 13, EEA, 2010.

C.J.J. Eijgenraam. Optimal safety standards for dike-ring areas. CPBDiscussion Paper 62, CPB Netherlands Bureau for Economic Policy Analysis, 2006.

C.J.J. Eijgenraam. From optimal to practical safety standards for dike-ring areas. *Water Science and Technology*, 56(4):113–124, 2007.

A. Elsner, S. Mai, and C. Zimmermann. Risikoanalyse – ein Element des Küstenzonenmanagements. *Coastline Reports*, 1:137–147, 2004.

EMOVE Project Consortium. Work Package 3 - European Joint Fact Finding. Part 1: Problem Identification. Technical report, RWS, MOW, BAW, COWI, Middelburg, 2015a.

EMOVE Project Consortium. Work Package 3 - European Joint Fact Finding. Part 2: Description of State, Impacts and Solutions. Technical report, RWS, MOW, BAW, COWI, Middelburg, 2015b.

[Engineering Committee for North and Baltic Sea] Engineering Committee. Die Sturmflut vom 16./17. Februar 1962 im niedersächsischen Küstengebiet. *Die Küste*, 10(1):17–54, 1962.

FGG Weser. EG-Wasserrahmenrichtlinie für die Flussgebietseinheit Weser. Bewirtschaftungsplan 2009 für die Flussgebietseinheit Weser. Maßnahmenprogramm 2009 für die Flussgebietseinheit Weser. Technical report, 2009.

C. Freitag, B. Hochfeld, and N. Ohle. Lebensraum Tideelbe. *Coastline Reports*, 9:69–79, 2007.

H. Giszas. Sturmflutschutz: Herausforderungen und Sicherheitskonzept. In *Proc. of HTG – Flüsse, Kanäle, Häfen, Sept. 2003*, pages 241–258. HTG – Hafenbautechnische Gesellschaft, 2003.

B. Gouldby and P. Samuels. Language of Risk. Project Definitions. Technical Report of the project FLOOD*site* - Integrated Flood Risk Analysis and Management Methodologies T32-04-01, HR Wallingford, 2005.

K.J. Gregory. The human role in changing river channels. *Geomorphology*, 79(3-4):172–191, 2006.

IAHR. Scheldt Estuary - Physics and integrated management. Technical report, Rijkswaterstaat, 2015.

[Intergovernmental Panel on Climate Change] IPCC. Adaptive options and Policy. Implications of Sea Level Rise and other Coastal Impacts of Global Climate Change. Workshop Report to the Coastal Zone Management Subgroup. Workshop Report, 1989.

[Intergovernmental Panel on Climate Change] IPCC. Climate Change. The IPCC Response Strategies. Report of Working Group III of First Assessment Report (FAR). Technical report, 1990a.

[Intergovernmental Panel on Climate Change] IPCC. Strategies for adaption to sea level rise. Technical report, 1990b.

[Intergovernmental Panel on Climate Change] IPCC. Global Climate Change and the Rising Challenge of the Sea. Technical report, 1992.

[Intergovernmental Panel on Climate Change] IPCC. Climate Change 1995: Impacts Adaption and Mitigation of Climate Change: Scientific-Technical Analysis. Contribution of Working Group II to the Second Assessment Report (SAR). Technical report, 1996.

[Intergovernmental Panel on Climate Change] IPCC. Climate Change 2013: The Physical Science Basis. Contribution of Working Group I to the Fifth Assessment Report (AR5) of the Intergovernmental Panel on Climate Change. Technical report, Cambridge University Press, 2013.

[Intergovernmental Panel on Climate Change] IPCC. Climate Change 2014: Mitigation of Climate Change. Contribution of Working Group III to the Fifth Assessment Report (AR5) of the Intergovernmental Panel on Climate Change. Technical report, Cambridge University Press, 2014a.

[Intergovernmental Panel on Climate Change] IPCC. Climate Change 2014: Impacts, Adaptation, and vulnerability. Part A: Global and Sectoral Aspects. Contribution of Working Group II to the Fifth Assessment Report (AR5) of the Intergovernmental Panel on Climate Change. Technical report, Cambridge University Press, 2014b.

[Intergovernmental Panel on Climate Change] IPCC. Climate Change 2014: Impacts, Adaptation, and vulnerability. Part B: Regional Aspects. Contribution of Working Group II to the Fifth Assessment Report (AR5) of the Intergovernmental Panel on Climate Change. Technical report, Cambridge University Press, 2014c.

C. Jeuken, D. Hordijk, C. Kuijper, P. Peeters, B. de Sonneville, and J. Vanlede. Koploperproject LTV-O&M – Thema Veiligheid - deelproject 1. Inventarisatie historische ontwikkeling van de hoogwaterstanden in het Schelde estuarium. Technical report, Delft, 2013.

M.C.J.L. Jeuken and Z.B. Wang. Impact of dredging and dumping on the stability of ebb-flood channel systems. *Coastal Engineering*, 57:553–566, 2010. doi: 10.1016/j.coastaleng.2009.12.004.

G. Kaiser. *Risk and vulnerability analysis to coastal hazards – an approach to integrated assessment.* Dissertation, University Kiel, 2006.

G. Kaiser, S. Reese, H. Sterr, and H.-J. Markau. ComRisk - Common strategies to reduce the risk of storm floods in coastal lowlands. Subproject 3: Public perception of coastal flood defence and participation in coastal flood defence planning. Technical Report, 2004.

R.J.T. Klein and R.J. Nicholls. Assessment of coastal vulnerability to climate change. *Ambio*, 28(2): 182–187, 1999.

A. Kortenhaus. *Probabilistische Bemessung von Seedeichen*. Dissertation, Technical University Braunschweig, 2003.

A. Kortenhaus, P. Fröhle, J. Jensen, N. von Lieberman, S. Mai, C. Miller, K. Peters, and H. Schüttrumpf. Probabilistische Bemessung von Bauwerken. *HANSA*, 144(4):68–76, 2007.

J. Kramer. Entwicklung der Deichbautechnik an der Nordseeküste. In J. Kramer and H. Rohde, editors, *Historischer Küstenschutz*, pages 63–109. Verlag Konrad Wittwer, Stuttgart, 1992.

H. Kunz. Salinity and water levels in the Weser estuary during the last hundred years – Anthropogenic influences on the coastal environment. In *Coastal Engineering*, pages 3533–3547, Kobe, 1995. ASCE.

H. Kunz. Sicherheitsphilosophie für den Küstenschutz. *Jahrbuch der Hafenbautechnischen Gesellschaft*, 54:253–287, 2004.

D. Lange, H. Müller, F. Piechota, and R. Schubert. The Weser Estuary. *Die Küste*, 74:275–287, 2008.

S. Mai. *Klimafolgenanalyse und Risiko für eine Küstenzone am Beispiel der Jade-Weser Region*. Dissertation, University Hannover, 2004.

H.-J. Markau. *Risikobetrachtung von Naturgefahren. Analyse, Bewertung und Management des Risikos von Naturgefahren am Beispiel der sturmflugefährdeten Küstenniederungen Schleswig-Holsteins*. Dissertation, University Kiel, 2003.

[Ministerie van Infrastuctuur en Milieu and Ministerie van Economische Zaken] MinInfraM and MinEZ. Delta Programme 2014. Working on the delta. Promising solutions for tasking and ambitions. Technical Report, MinInfraM and MinEZ, 2013.

[Ministerie van Infrastuctuur en Milieu and Ministerie van Economische Zaken] MinInfraM and MinEZ. Delta Programme 2015. Working on the delta. The decision to keep the Netherlands safe and liveable. Technical Report, MinInfraM and MinEZ, 2014a.

[Ministerie van Infrastuctuur en Milieu and Ministerie van Economische Zaken] MinInfraM and MinEZ. Draft National Water Plan (2016-2021). Technical Report, MinInfraM and MinEZ, 2014b.

[Ministerie van Infrastuctuur en Milieu and Ministerie van Economische Zaken] MinInfraM and MinEZ. Delta Programme 2016. Working on the delta. And now we're starting for real. Technical Report, MinInfraM and MinEZ, 2015.

G. Mol. De Westerschelde: een resultaat van menselijke ingrepen. Technical Report RIKZ-95.030, Rijksinstituut voor Kust en Zee (RIKZ), 1995.

H. Oumeraci and A. Kortenhaus. Risk-based Design of Coastal Flood Defences: A Suggestion for a Conceptual Framework. *Proc. ICCE 2002*, 2:2399–2411, 2002.

[Numerous Authors] Parry et al. Technical Summary. In *Climate Change 2007: Impacts, Adaptation and Vulnerability. Contribution of Working Group II to the Fourth Assessment Report (AR4) of the Intergovernmental Panel on Climate Change*, pages 23–78. Cambridge University Press, 2007.

E.B. Peerbolte, J.J. de Ronde, L.P.M. de Vrees, M. Mann, and G. Baarse. *Impacts of Sea Level Rise on Dutch Society, a description of the ISOS (=Impact of Sea Level Rise on Society) case study for the Netherlands*. Delft Hydraulics and Rijkswatertstaat, Delft and The Hague, The Netherlands, 1991.

H.P. Peters and H. Heinrichs. Das öffentliche Konstrukt der Risiken durch Sturmfluten und Klimawandel. In *Land unter? Klimawandel, Küstenschutz und Risikomanagement in Nordwestdeutschland: die Perspektive 2050*, pages 115–144. Oekom Verlag, 2007.

S.T. Plapp. Perception and Evaluation of Natural Risks – Interim report on first results of a survey in six districts in Germany. Working Paper 1, University Karlsruhe, 2001.

S.T. Plapp. *Wahrnehmung von Risiken aus Naturkatastrophen. Eine empirische Untersuchung in sechs gefährdeten Gebieten Süd- und Westdeutschland.* Dissertation, University Karlsruhe, 2003.

O. Renn and A. Klinke. A framework of adaptive risk governance for urban planning. *Sustainability*, 5(5):2036–2059, 2013. doi: 10.3390/su5052036.

O. Renn, R. Carius, H. Kastenholz, and M. Schulze. Erik – entwicklung eines mehrstufigen verfahrens der risikokommunikation. Technical Report, 2005.

[Rijksinstituut voor Kust en Zee] RIKZ. Waterbewustzijn in Nederland: Leren van risicobewustwordingsprocessen in het buitenland. RIKZ/2003.005, RIKZ, BWD and ERGO, 2002.

[Rijkswaterstaat] RWS. National Water Plan (2009-2015). Technical Report, RWS, 2009.

[Rijkswaterstaat] RWS. Room for the River. Safety for million people in the Dutch delta. Technical Report, RWS, 2012.

[Rijkswaterstaat] RWS. Tailor-made Collaboration. A clever combination of process and content (Room for the River). Technical Report, RWS, 2013.

J. Schanze. Preliminary conceptual framework of integration ("Dresden Paper"). Technical Report, IOER, 2006.

B. Schuchardt and M. Schirmer. *Land unter? Klimawandel, Küstenschutz und Risikomanagement in Nordwestdeutschland: die Perspektive 2050*. Oekom Verlag, 2007.

H. Sterr, R.J.T. Klein, and S. Reese. Climate Change and the Coastal Zone. An overview of the state-of-the-art on regional and local vulnerability assessment. Working paper 38.2000, FEEM, 2000.

[Technical Advisory Committee on Water Defence] TAW. Fundamentals on Water Defence. Technical report, TAW, 1998.

TIDE Project Consortium. Management of Estuaries. The need to understand nature and society. TIDE summary report. Technical report, Hamburg Port Authority (HPA), Hamburg, 2013. URL www.tide-toolbox.eu.

I. Umlauf, J. Meyerdirks, and F. Ahlhorn. Identifikation zukünftiger Konfliktpotenzial im niedersächsischen Küstenbereich im Projekt "Speichern statt pumpen". *Coastline Reports*, 17: 191–202, 2011.

W. Vandenbruwaene, Y. Plancke, T. Verwaest, and F. Mostaert. Interestuarine comparison: hydrogeomorphology: hydro- and geomorphodynamics of the TIDE estuaries Scheldt, Elbe, Weser and Humber. Technical report, flanders Hydraulics Research, 2013. URL www.tide-toolbox.eu.

R. Vergouwe. The National Flood Risk Analysis for The Netherlands. Technical report, Ministrie van Vekeer en Waterstaat, Interprovincial Overleg, Dutch Water Authorities, 2015.

[Vlaams Nederlandse Schelde Commissie] VNSC. Water, zand en slib in het Schelde-estuarium. Inzichten uit het Vlaams-Nederlandse onderzoek. Technical report, 2013.

[Vlaams Nederlandse Schelde Commissie] VNSC. T2009-Rapport Schelde Estuarium. Technical Report C03041.002718.0400, 2014.

Z.B. Wang, M.C.J.L. Jeuken, H. Gerritsen, H.J. de Vriend, and B.A. Kornman. Morphology and asymmetry of the vertical tide in the Westerschelde estuary. *Continental Shelf Research*, 22: 2599–2609, 2002.

[German Advisory Council on Global Change] WBGU. *World in Transition – Strategies for Managing Global Environmental Risks. Executive Summary*. WBGU, 1999. ISBN 3-9806309-3-5.

[World Commission on Environment and Development] WCED. *Our Common Future. World Commission on Environment and Development/WCED (Brundtlan's Commission).* Oxford University Press, UK, 1987.

J.C. Winterwerp. On the response of tidal rivers to deepening and narrowing. Technical report, MOW, VNSC, RWS, 2013.

Prospects: Methodologies of Integration 7

7.1 A Project Process Cycle

The intention of integrated coastal zone management is the integration of the variety of interests and needs which can be found at the coast. The way on how to achieve a sustainable coastal development might be the establishment of a continuous and adaptive process. Despite this goal ICZM is mainly conducted as project either inter- or multidisciplinary.[1] The amount of literature on project management and planning is awesome, here, selected references are introduced to exemplify basic aspects of project initiation and management. One has to consider that each project is unique even though you are already experienced because a variety of uncertainties have to be managed.

In Fig. 7.1 a generalized four-step project process is drawn based on Baguley (1995) and GESAMP (1996). The four steps could be described as follows:

Step 1 – Analysis Herein the basic preparatory steps for the project have to be taken such as the identification of the respective coastal system, the policy and institutional context as well as the development context. The first point is targeting on the description of the natural environment interlinked with the issue of concern. The second point requires an investigation of the respective governance structures and arrangements. The latter point touches the identification of relevant stakeholders and the social perception of the issue of concern. Main task is to compile, to integrate, to assess and to prioritize the available information and data.

[1] The distinction between inter- and multidisciplinary is that in interdisciplinary projects different research fields or sectors work together to achieve a certain goal. In multidisciplinary projects different research fields or sectors are working together on the same issue but not necessarily in close cooperation.

F. Ahlhorn, *Integrated Coastal Zone Management*,
Wasser: Ökologie und Bewirtschaftung, DOI 10.1007/978-3-658-17052-3_7

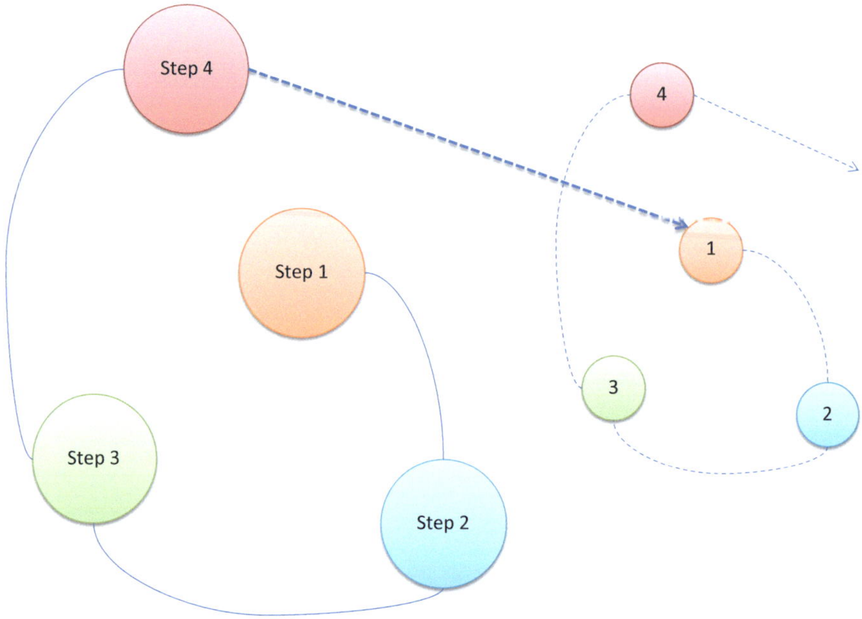

Fig. 7.1 Generalized steps of a project process. Source: based on Baguley (1995) and adapted from GESAMP (1996)

Step 2 – Design Based on the results of step 1 the final project plan and structure should be developed. For example, competences and responsibilities is determined for the internal project management and for external project communication. Furthermore, the participation process has to be designed based on the results of the stakeholder analysis conducted in step 1. Performance and stop criteria are developed and jointly agreed by the project consortium.

Step 3 – Implementation Within this step the work has to be conducted which has been planned and scheduled in the previous steps. The project manager (or management group) has to steer the process to achieve a successful project. Helpful are the permanent observation and periodically assessment of the performance and stop criteria formulated in step 2.

Step 4 – Conclusion and finish The last step contains the phaseout part of the project. In this part the project achievement could be evaluated by internal as well as external project participants. For example, the experiences gained during the process should be documented and act as knowledge base for the next project. Furthermore, open questions or new tasks emerging during project lifetime can be transformed into a new project. Consequently, a new step 1 starts a new project process cycle.

7.2 Participatory Action

7.2.1 Introduction

Why do people still worry about the status of nature or the natural environment, for example, the coastal environment, but usually act such as their behavior don't cause impacts? The evolvement of ICZM over the last decades shows a similar picture as drawn for the environmentalism. Most of the people are worrying about the bad status and deterioration of natural assets, but show a contradictory behavior. The reasons for that behavior are diverse and won't be in the focus of this section. Nevertheless, it seems to be important to find new ways on how to deal with existing apprehension and dissatisfaction of both the current status and anticipated future development of the natural environment and the process and the enforcement of environmental legislation (e.g., Renn and Webler, 1994; Renn, 1999).

The traditional policy approach of *decide-announce-defend* (DAD, Fig. 7.2) is no longer accepted by a variety of interest groups or parts of the general public. Therefore, over the course of time more participatory ways of (policy) decision-making processes evolved. For example, in Sect. 2.1 the beginning of ICZM in the USA (San Francisco Bay) indicates a new way of dealing with coastal (environmental) problems (see also WFD explanations). Consequently, a more collaborative way for decision-making has been developed, so that the traditional DAD approach is supplemented to the *engage-deliberate-decide* approach (EDD, Fig. 7.2). For example, Renn and Webler (1994) and Renn (2006) discussed why more deliberative approaches were necessary, especially for decisions concerning environmental and risk aspects.

In general, the way of decision-making has been amended by the active involvement of stakeholders. The definition of the term *stakeholders* should be seen as broad as possible in the first stage:

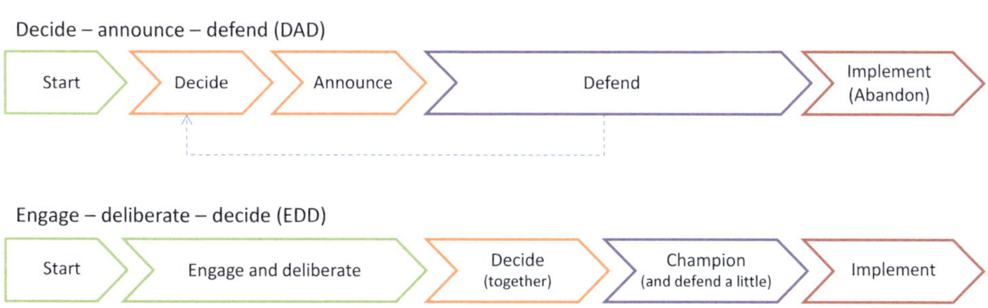

Fig. 7.2 Decide-announce-defend (DAD) versus Engage-deliberate-decide (EDD) as processes for decisions. Source: adapted from Walker (2009)

> Stakeholders are all institutions, organizations and people that have a stake at the issue of concern.

In literature you can find a variety of explanations and definitions for active participation of stakeholders in decision-making processes such as collaborative planning, co-management or alternative dispute resolution and many more. Here, the definition of the term *collaborative governance* as proposed in Agger and Löfgren (2008, p. 544) is used to define participatory action:

> A governing arrangement where one or more public agencies directly engage non-state stakeholders in a collective decision-making process that is formal, consensus-oriented, and deliberative and that aims to make or implement public or manage public programs or assets.

Ansell and Gash (2008) argue why this definition is appropriate in most cases. Important is that the definition already contained six relevant criteria for the distinction between various collaborative (participatory) approaches.

Figure 7.2 visualizes the main steps of the DAD approach. The length of the boxes indicate the time span of each step. In the DAD approach the first two steps are not that time consuming, because the respective agency or the group of people are jointly agreeing on the decision, for example, a law or an enactment. These steps were more or less executed without public involvement, so that the decision is published afterwards for the first time to a wider audience. For example, the formal procedure of a planning approval process requires a hearing after the publication of the completed plan. Hence, only petitions and complains can be submitted. These petitions have to be processed, that is, compiled and assessed if and how they will be taken into account and if and how they may lead to a revision or amendment of the original plan. Sometimes the original plan has to be rejected (DAD step 5, Fig. 7.2).

The phases of the EDD approach are different. The largest time is allocated to the deliberation phase when – ideally – all relevant stakeholders (actively) participate in a deliberation and negotiation phase to find the best solution for the given problem or task. Either this solution might be a compromise or something else, important is that the group tries to approach consensus as much as possible (EDD Step 3, Fig. 7.2). The publication of the joint decision might went smooth if all relevant stakeholders participated in the previous steps and none of them followed a hidden agenda to jeopardize the decision afterwards.

To confine multifaceted ways of organizing and conducting a collaborative process the following section will provide some criteria. Important to keep in mind is that each participatory process has to be tailor made and adapted to the current context you are working in.

7.2.2 Basic Criteria

Based on the analyses of 137 case studies Ansell and Gash (2008) developed a general model for collaborative governance (Fig. 7.3). Indication for a starting point of a

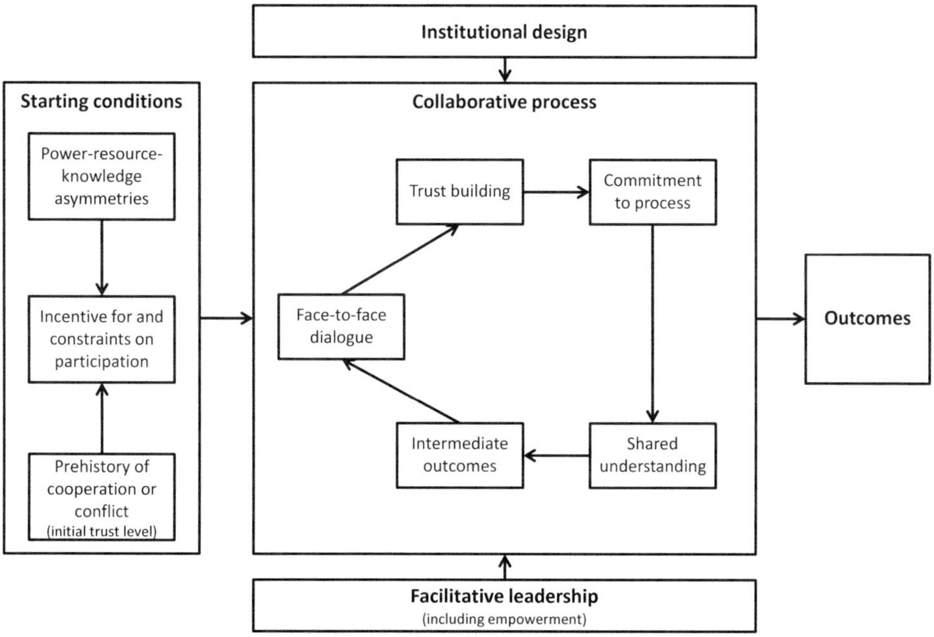

Fig. 7.3 The collaborative governance model deduced analyzing a variety of case studies. Source: adapted from Ansell and Gash (2008)

collaborative process is given on the right-hand side of the diagram. For example, Renn and Webler (1994, p. 13–15) identified why a collaborative process could be initiated: equity, uncertainty on the distribution of environmental risks, public perception of controversial expert dialogs, controversial on the scale of risk assessments between experts and public, appropriation of the living environment by bureaucrats and technical environmental changes, and protest against the decision-making process.

The collaborative process itself consists of five steps. First, the face-to-face dialog with the identified stakeholders to build trust and investigate their expectations and apprehensions (second). Third, all participating parties have to be committed to the rules and the goals of the process. In the fourth step (shared understanding) the entire group jointly agrees on a clear mission, the identification of the problem and common values for the process. Fifth, intermediate outcomes will spread and intensively discussed in the group, so that a step-by-step approach will lead to the final outcome of the collaborative process. This process is taking place in a specific institutional framework (institutional design) which has to be announced transparently and traceable for every member of the group. In many processes the leadership of the process is given to an external expert or group of experts which have to be acknowledged by all parties involved.

The identification of stakeholders and the conduction of a force-field analysis (see, e.g., Lewin and Cartwright 1951; Tucker 1979; Thomas 1985) is mainly based on literature review and internet research. The force-field analysis provides a first glance on the

Resource	Predominant Sector	Medium/Tool	Motivation
Money	Economy	Capital transfer	Economic incentives
Power	Politics	Authority, force	Fear for punishment
Social prestige	Social system	Reputation, award	Solidarity, support
Cultural accountability	Culture	Belief, sense	Trust, snugness
Evidence	Science	Proof, truth	Insight in expected consequences

Fig. 7.4 Societal resources and their impacts in society. Source: adapted from Renn and Webler (1994)

interrelationships of stakeholders and stakeholder groups which is necessary information for the designing phase of the project and, especially, the face-to-face interviews. But, if the stakeholders are identified how to understand why they are motivated to be part of collaborative process?

Renn and Webler (1994) proposed a framework on societal resources to assess why stakeholders might attend a participation process. Each societal resource is described by the predominant sector it belongs to and the tools which are used to apply the respective resource to achieve certain goals (Fig. 7.4). The basic assumption is that stakeholders can achieve their goals best if they possess many societal resources. Five different societal resources are proposed:

Money is mainly linked to the economic sector which offers power and influence for the respective actor or interest group.

Power is mainly linked to the political sphere which offers the ability of steering and regulating societal activities.[2]

Social prestige in society is a dominant motivation for social action, expressed, for example, through a prize or honor.

Cultural accountability is important as far as the affiliation to a specific political or societal group is concerned. Opinions are influenced by cultural values or a philosophy of life.

Evidence is the ability to prove the consequences of one's own or others action. Today this ability is getting more and more important, see, for example, the intention of the implementation of the IPCC (see Sect. 6.1.2).

[2] *Power* is the possibility to force somebody to do something also against his will. Definition according to Weber (1922, §16)

If the motivation and reason why stakeholders take part of the process is clear it is now important to know which requirements a collaborative process has to fulfill. The following list of criteria is a choice taken from Linder and Vatter (1996, p. 182–188):

Equity A participatory process should strive to reduce as much as possible the disparity (according to the access to knowledge and societal capability of attending such a process) between the attending parties.[3] For example, the broad public should have the opportunity to participate in informal pre-project meetings. The differences in existing and non-existing power in specific groups should be minimized within such processes, that is, try to involve also powerless interest groups or people which are mostly excluded for different reasons, for example, disadvantaged through less financial support.

Transparency A transparent and traceable process is important to build trust between the attending parties in a participation process. The open discussion on values, interests and needs can help to improve the quality of discussion within the process.

Capability for social learning Not only scientific knowledge can help to find best solutions for a given problem or challenge also local knowledge provided by "layman" can provide insights in specific local circumstances and contexts (see, for example, Bormann et al., 2012).

Early and iterative involvement To involve stakeholders from the beginning is important because problem definition and agreements on the project procedures will set in first phases of a participation project. The earlier people are involved to bring own views and opinions the more committed they feel for the entire process. Furthermore, an iterative process offers the opportunity to involve stakeholders which were not identified in the first phase of the project either they are not known or their attendance is necessary caused by the project status (see, for example, Bormann et al., 2015).

Intelligible information and clear conflict resolution rules The accessibility and comprehension of available information should not be limited to selected participants because it can raise mistrust and aversion against the process. The fear that an open (informal) process lasts longer than formal procedure has not been proven by reality – see, for example, Würtenberger (1993) and Walker (2009).

Joint agreement on a code of practice and decision-making process Important is to achieve consensus on the code of practice and the decision-making process within the

[3] Linder and Vatter (1996) proposed also the criteria *competency of participants*, *social compensation* and *compensation between powerful and powerless parties*, here these will be subsumed under the criterion *equity*.

participation process, so that every participant is clear about the rules, also in case of a dispute.

Commitment The degree of compulsion of the collaborative (decision-making, governance) process has to be as clear as possible from the beginning. For the further process phases and subsequent implementation phase it is important for the stakeholders to know how binding the jointly developed results are.

Motivation of participants The motivation of stakeholders might be influenced by a variety of reasons and circumstances, the description of the societal resources might help as guideline.

Compensation between short- and long-term interests The process should take care of the ability to balance the interests of some stakeholders or stakeholder groups to strive for short-term solution of problems (own interest) and the desired long-term effects of broader solution approaches (social implications, e.g., taking into account future generations as for sustainable development).

7.2.3 Principal Process

Based on the previous parts a principal process for collaborative action can be drawn (Fig. 7.5). The process is divided into four steps, each has to be adjusted to the specific situation and context the process should be applied to.

In general, the first step should be the analysis of the situation/context the task and subsequent the identification of all relevant stakeholders. A force-field analysis is helpful to identify the power and forces between the various stakeholders and stakeholder groups. The first step might be finished by the compilation of this information and data. Important to reflect is that the result of a stakeholder analysis indicates the affiliation of persons to specific groups of interest. According to Sabatier (1998) these affiliated persons must not necessarily share or represent the interests and needs of the respective stakeholder group in total. For example, scientists might not be fully neutral in a natural resource management

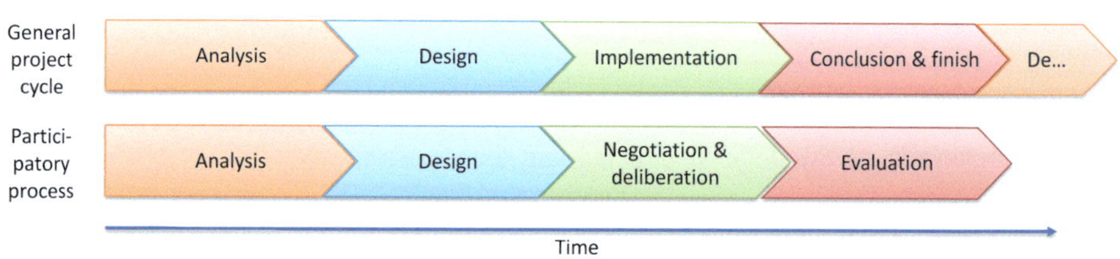

Fig. 7.5 Principal steps of a participatory process

process or persons might have different beliefs than they have to represent in their role as agency officials. Thus, the arrangement of stakeholders should not solely follow the primordial affiliation to a specific interest group or agency, but also try to take the beliefs and interest into account.

Within the second step the compiled information is used to design the participation process and to finally plan the necessary steps to achieve the desired goal(s). In this step it is helpful to approach the stakeholders directly because of insufficient information base or the wish to deepen the knowledge on interests and needs of all or selected stakeholders. A helpful tool is to draw a communication plan on how to disseminate and communicate inside and outside the process.

The negotiation and deliberation phase is the centerpiece of the collaborative process. The process plan, the rules and procedures have to be presented and discussed, and – ideally – consensus has to be achieved by the group on these aspects. Afterwards, either a revision of the plan is necessary or not. In this step a variety of tools and methods can be applied to achieve the goal(s) of the entire process. For example, Wittmer et al. (2006) proposed specific criteria on how to decide appropriate tools. Further information could be found, for example, in Sabatier (1998); Borrini-Feyerabend et al. (2000); Petts and Leach (2000); OECD (2001); Wenzel (2001); Delbridge et al. (2002); Drouet et al. (2002); van Asselt Marjolein and Rijkens-Klomp (2002); Numerous Authors (2006) and Salter et al. (2010)

Subsequently, the participation process should be finished by an evaluation step. Herein, the participants should be motivated to assess the entire process and provide ideas of improvement for next processes.

7.3 Assessment Schemes

7.3.1 Introduction

Starting point of evaluation methods are the preferences of individuals and organizations. The second aspect is the maximization of benefits. To quantify the preferences and the maximization of benefit, it is mandatory to carry out an "economic appraisal" to evaluate costs and benefits of a project (cost-benefit analysis, CBA). This is indispensable for the private as well as for the public sector, because the acceptance decreases to spend tax on inefficient projects. The difference between the private sector and the public sector is that cost and beneficial aspects are considered in a wider sense. That means the entire effects of the project should be considered, not limited to one company.

First approaches to determine costs and benefits in water management have been developed in Germany in the 1980s – see, for example, LAWA (1981) and DVWK (1985). Related investigations which were conducted regarding coastal protection projects were:

Benefit Analysis of the Special Plan Coastal Protection for Sylt by Klaus (1986), *Economic Appraisal for the Potential Flood prone Area in Schleswig-Holstein* by Klug et al. (1998), *Sea Level Rise and Socio-Economic Consequences* by Behnen (2000) and the investigation about micro scale evaluation of risk of flood prone coastal zones by Reese et al. (2002). Only Klaus (1986) conducted a CBA, the other investigations focus on different items, which contributes to a CBA. For water management and coastal protection in Germany to some extent utility analysis and cost-effectiveness analysis were conducted, but are met with skepticism (Hartje et al., 2002).

The fundamental point of CBA is to try to make assessments transparent and fit for decisions because most decisions in our world are mainly taken on the basis of economic benefit. That means that evaluation methods should contribute to demonstrate the relevance of non-economical aspects, for example, biosphere or nature (WBGU, 1999). Therefore, the aim of such methods is not to provide an exact figure, but to strive for the integration of all value categories as shown in Fig. 7.6. WBGU (1999) stressed that a focus should lie on the integration of economic evaluation, societal decision-making and the aims of sustainability. The application of a CBA in environmental projects is under discussion – see, for example, Hartje et al. (2002); Hansjürgens (2004); Ruijgrok (2005); Antunes et al. (2006); Convery (2007) and Barbier et al. (2011). Three main point of criticism about CBA exists (Hansjürgens, 2004, p. 246):

- Criticism of the efficiency consideration as the underlying normative approach. This criticism is fundamental by nature and totally rejects CBA on the basis of ethical considerations.
- Criticism of insufficient specifications of CBA. This criticism is directed at methodological shortcomings of CBA.
- Criticism of using CBA in the political decision-making process.

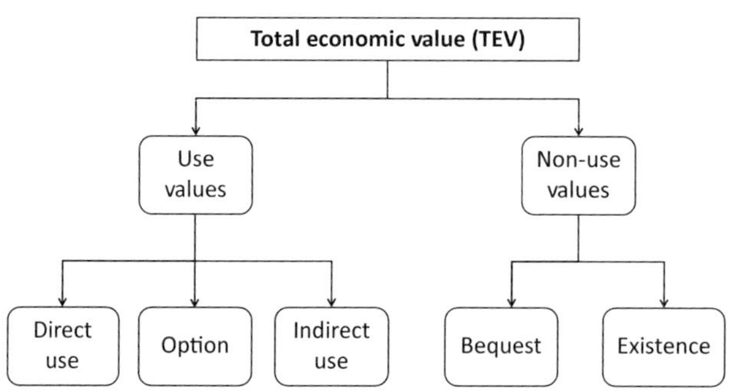

Fig. 7.6 Total economic value (TEV) approach. Source: modified from Barbier (1990) and WBGU (1999)

The first point focuses on the weighting of criteria within a CBA. The weighting of criteria should discriminate one criteria against another, but decision in real world are made by weighting the considered criteria. The second one focuses on the uncertainty, the arbitrariness and the feigned accuracy of data. Strong effort has to spend on the issue of data collection and data processing. There is no universal solution for these issues, but the applied method should be transparent regarding the implementation of data, the uncertainty and the accuracy. In most projects is less time for comprehensive data collection and verification. The last point of criticism is focused on the preference of politicians to decide on *hard* rather than on *soft* facts. Aspects which can be monetarized are seen as hard facts, the others as soft facts. Hansjürgens (2004) stated that the critic is related to the incorrect use of the outcomes of a CBA rather than to the method itself.

The problem related to the issue of discounting is seen as a limitation to the application of CBA. Discounting should take cost and benefit of different time periods into account to enable a comparison (Hampicke, 1991). Economists indicate that discounting should be applied for periods not longer than 10–20 years because this period includes the time scale of one generation where the effects may be estimated, but to assess the effects which might occur for the next generations is impossible (regarding cost and benefit, see, e.g., Hampicke 1992; Endres and Holm-Müller 1998).

Conflicts may arise taking societal, economical and ecological aspects into account in a CBA. Knowing the different points of view it is possible to anticipate potential conflict reasons by finding new solutions and compromises. Mainly, the applied methods of CBA try to monaterize goods which value cannot be determined by, for example, market price. Monetary assessment is mainly done by willingness-to-pay methods. Different methods exist which can be used trying to investigate how much the customers or users are willing to pay, especially for values which cannot be directly quantified (non-tangible). The advantages of willingness-to-pay methods are that goods are related to a calculated value and that they can be integrated in a CBA. The main disadvantage of willingness-to-pay methods is that these values represent not really the price people would pay for the good – see, for example, Bateman (1995); Breidert (2005) and Guzman and Kolstad (2007). To estimate the influence of certain aspects regarding the statements in willingness-to-pay investigations different methods can be applied.

Figure 7.6 outlines the approach of the total economic value (TEV, Barbier 1990; Pearce and Turner 1990) which tries to cover all relevant value categories for a comprehensive appraisal. The approach considers different spatial dimensions (e.g., nearby or far away) as well as temporal dimensions (e.g., short-term or long-term) and it is possible to demonstrate the total amount of values which are related to the demand of natural goods. The aim of this approach is not to allocate one calculated figure, but to provide a discussion platform which considers all relevant value categories (e.g., benefit value, functional value, existence value, etc.). This comprises explicitly the ecological and social services of the biosphere, and, for example, the natural coastal environment.

Selected value categories are described to get an impression on how to understand the classification:

Benefit value describes the direct use of services of the biosphere for production or consumption. The value of experience is also subsumed under this value category, because it is related to consumption, that is, tourism.

Functional value describes the indirect services of the biosphere. The ecological systems provide many regulation functions (e.g., water cycle, biogeochemical cycles or the composition of gases in the atmosphere) and structures (e.g., soil, dunes or slopes), which are necessary for human survival on the earth. Functions are usable characteristics of ecological structures. Values for the ecological goods consists of the utilization of provided functions. The functional value describes the quality or the actual state of usable ecosystem services. For example: a wide fore land provides a regulation function for coastal protection, because it reduces to some extend the hydrodynamical load on the main dike, the soil provides a production function for agricultural use and the existence of habitats, species and natural processes provides an information function for nature conservation.

Existence value is a non-use value. The existence value is neither based on direct nor indirect utilization, only the knowledge of the existence of natural goods provide a value. In particular, the existence value is related to natural goods which will never be used even in the future. For example, to make the existence value tangible a questionnaire could be applied to investigate the characteristics of the existence value of the stakeholders for a specific region.

Consequently, the requirements for a socio-economic and ecological evaluation scheme which should be used and applied are as follows: the method should

- integrate socio-economic as well as ecological aspects,
- enable to evaluate tangible as well as in-tangible aspects,
- integrate monetary as well as non-monetary values,
- be able to consider and to compare categorized criteria of different aspects (multi criteria analysis),
- be able to compare different scenarios and
- be transparent and traceable.

CBA are mainly conducted by using a utility function which should be either maximized or minimized. Thus, the aim is to find an optimal solution for problems. Most real-world problems are not related to one criterion, but to many criteria like social, economical and ecological criteria. Applying a utility function it is impossible, for example, to maximize all criteria. On the other hand, in real-world problems it is not wise to maximize all criteria it is better to find a compromise. Therefore, a new category of multicriteria decision-aiding methods has been developed: outranking methods

(Brans and Vincke, 1985; Roy, 1985, 1996). As an example, the outranking method *P*reference *R*anking *O*rganisation *Meth*od for *E*nrichment *E*valuation (PROMETHEE) is choose here.

7.3.2 Multicriteria Decision Aid Methods: An Outline

The objective of outranking methods is to support the decision maker. Outranking methods do not focus on maximizing or minimizing a utility function, their intention is the pairwise comparison of criteria of two alternatives. Outranking methods use preference functions which indicate the preference, indifference or incomparability of criteria. A comprehensive overview about outranking methods and their characteristics can be found in Guitouni and Martel (1998) and a more general and comprehensive description of multiple criteria decision analysis (MCDA) can be found in, for example, Belton and Stewart (2002) and DCLG (2009). Two outranking methods were developed in the earliest stage of MCDA and now are widely used for various purposes: ELECTRE and PROMETHEE. ELECTRE (Elimination et Choice Translation Reality) was applied to many problems to support decision makers, but has the disadvantages to require a lot of parameters. Some of them influence the results in a way which were not clearly understood and the application was more complex than others (Brans and Vincke, 1985; Belton and Stewart, 2002; Ruhland, 2004). Salminen et al. (1998) compared three different outranking methods in the context of environmental problems and came to the conclusions that more than one method should be applied if possible. The differences are not great between the outcomes of the different methods, they vary in their complexity.

Guitouni and Martel (1998) provided tentative guidelines to chose the right method:

Guideline 1 Determine the stakeholders of the decision process. If there are many decision makers, one should think about group decision-making methods or group decision support systems.

Guideline 2 Consider the decision maker "cognition" when choosing a particular preference elucidation mode.

Guideline 3 Determine the decision problematic pursued by the decision maker. If the decision maker wants to get an alternatives ranking, then a ranking method is appropriate, and so on.

Guideline 4 Choose the multicriterion aggregation procedure (MCAP) that can handle properly the input information available and for which the decision maker can easily provide the required information; the quality and the quantities of the information are major factors in the choice of the method.

Guideline 5 The compensation degree of the MCAP method is an important aspect to consider and to explain to the decision maker. If he refuses any compensation, then many MCAP will not be considered.

Guideline 6 The fundamental hypotheses of the method are to be met (verified), otherwise one should choose another method.

Guideline 7 The decision support system coming with the method is an important aspect to be considered when the time comes to choose a multicriteria decision analysis method.

Brans and Mareschal (2005) emphasized in their article that PROMETHEE shall overcome the disadvantage of many evaluation methods to calculate an optimal solution using a utility function. Hunjak (1997) and Esser (2001) proved that the outranking method PROMETHEE is well founded in mathematics and fulfilled the demand of information enrichment. Furthermore, divers extension can be found for the outranking method PROMETHEE and others such as ELECTRE – see, for example, Hisschemöller et al. (2001); Marinoni (2005); Fernndez-Castro and Jimnez (2005) and Corrente et al. (2013).

7.3.3 Principal Process

Based on the previous section a principal process for the assessment in a participatory process can be outlined. Mostly, the problems in environmental or coastal management are multifaceted and, thus, demand for the integration of multiple criteria and indicators. A variety of sectors and different interests and needs are touched and have a stake in the given task. Consequently, the application of a single utility function is mostly not appropriate. Multicriteria decision aid methods provide the advantage that multiple criteria could be handled and assessed. The principal process shown in Fig. 7.7 indicates the basic steps which might be taken in the planning of an assessment process.

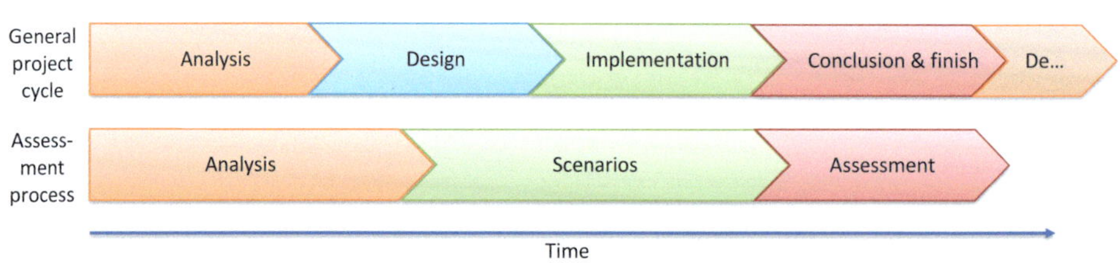

Fig. 7.7 Principal phases of an assessment process

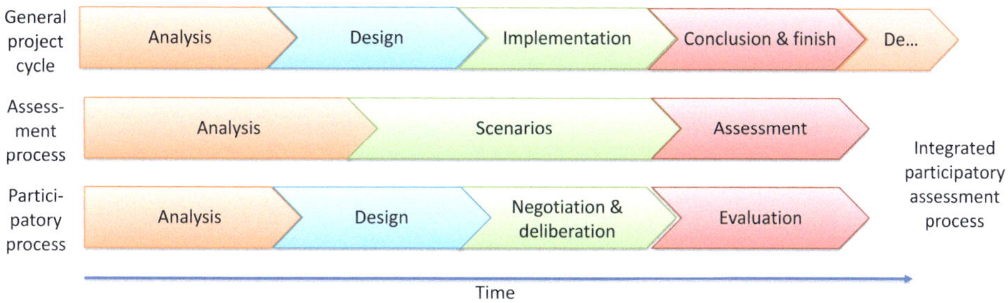

Fig. 7.8 Principal sketch of a participatory integrated assessment process

Planning and designing of the assessment process is done in the first step. The appropriate assessment method should be identified for the given problem. A first identification and compilation of indicators and criteria has to be conducted. The second step is the development of alternative options for the problem. The list of indicators and criteria should be assessed against the appropriateness to decide which option is best. Sometimes this list has to be amended or revised because necessary data is not available or the data quality is insufficient. The last step contains the assessment which of the identified options is appropriate to solve the problem. Here, multicriteria decision aid methods can help to find an optimal solution based on a variety of criteria. The advantage of the MCDA methods is that multiple criteria could be treated in a different way such as optimizing one criteria and minimizing another one.

7.4 Participatory Integrated Assessment

The conjunction of the participatory process and the assessment process previously outlined could be called a participatory integrated assessment process (PIA). Finally, the amalgamation of respective steps of each process to a joint procedure and to involve the stakeholders in each step of the process is important.

Many publications can be found which applied the PIA process to solve environmental or coastal management problems – see, for example, Georgopoulou et al. (1998); Wenzel (2001); Messner et al. (2003); Ahlhorn (2009); Grau et al. (2010); Hunag et al. (2011); Ortuno (2013); Malczewski and Rinner (2015); Martin et al. (2016) and Rebolledo et al. (2016).

It is important to consider that every participatory integrated process has to be adjusted to the respective context and situation, thus, a tailor-made approach is desirable. There is no blueprint available for such processes (Fig. 7.8).

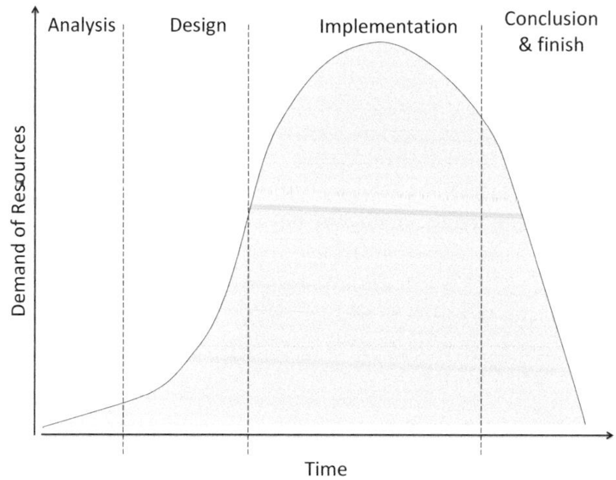

Fig. 7.9 Generalized project life cycle indicating the effort of resources in time. Source: adapted from Baguley (1995)

Exercise

1. In Sect. 7.2 it is opted for a collaborative governance process. Renn and Webler (1994) discuss the pros and cons of different decision-making processes and, finally, argue for a cooperative process involving stakeholders from the beginning. Try to comprehend the argumentation and discuss what are the benefits and what are the pitfalls – see, for example, Renn (1999, 2006) and Renn and Klinke (2013).
2. In Sect. 7.3 assessment schemes have briefly been outlined. Discuss the differences between economic utility functions and methods of multicriteria analysis. What are the pros and cons with regard to coastal management issues?
3. In Sect. 7.4 a principal participatory integrated assessment process is proposed as tool of choice. Now the proposed scheme should be enlivened for adjusted application to different tasks. Design a detailed participatory integrated assessment process using different participatory and assessment tools. See also Sect. 8.1.
4. Regarding the assessment of processes and projects the question may be raised how to assess progress in integrated coastal zone management? A variety of literature can be found dealing with the identification of appropriate indicators and criteria for the status and the progress of ICZM, here, only a few are selected, for example, Cendrero and Fischer (1997); Burbridge (1997); Olsen et al. (1999) and Olsen (2003). Based on these, and of course others, compile a list of indicators and criteria for the assessment of coastal zone management.
5. Figure 7.9 indicates the demand of resources in projects as well as in processes in general. Explain the diagram.

References

A. Agger and K. Löfgren. Democratic assessment of collaborative planning processes. *Planning Theory*, 7(2):145–164, 2008. doi: 10.1177/1473095208090432.

F. Ahlhorn. *Long-term Perspective in Coastal Zone Development. Multifunctional Coastal Protection Zones*. Springer, 2009.

C. Ansell and A. Gash. Collaborative governance in theory and practice. *Journal of Public Administration Research and Theory*, 18(4):543–571, 2008.

P. Antunes, R. Santos, and N. Videira. Participatory decision making for sustainable development – the use of mediated modelling techniques. *Land Use Policy*, 23(1):44–52, 2006. doi: 10.1016/j.landusepol.2004.08.014.

P. Baguley. *Managing successful projects: a handbook for every manager*. Pitman, 1995. ISBN 978-0-273-61377-0.

E.B. Barbier, S.D. Hacker, C. Kennedy, E.W. Koch, A.C. Stier, and B.R. Silliman. The value of estuarine and coastal ecosystem services. *Ecological Monographs*, 81(2):169–193, 2011. doi: 10.1890/10-1510.1.

E.D. Barbier. Natural resource degradation: Policy, economics and management. Technical Report Gatekeeper Series No. LEEC 90-01, 1990.

I. Bateman. Environmental and economic appraisal. In T. O'Riordan, editor, *Environmental Science for Environmental Management*, pages 45–65, 1995.

T. Behnen. *Der beschleunigte Meeresspiegelanstieg und seine sozio-ökonomischen Folgen: Eine Untersuchung der Ursachen, methodischen Ansätze und Konsequenzen unter besonderer Berücksichtigung Deutschlands*. Dissertation, University Hannover, 2000.

V. Belton and T.J. Stewart. *Multiple Criteria Decision Analysis*. Kluwer Academic Publishers, 2002.

H. Bormann, F. Ahlhorn, and T. Klenke. Adaptation of water management to regional climate change in a coastal region – hydrological change vs. community perception and strategies. *Journal of Hydrology*, 454-455:64–75, 2012.

H. Bormann, R. van der Krogt, L. Adriaanse, F. Ahlhorn, R. Akkermans, Y. Andersson-Sköld, C. Gerrard, N. Houtekamer, G. de Lange, A. Norrby, N. van Oostrom, and R. de Sutter. *Handbook of Climate Change Adaptation*, chapter Guiding Regional Climate Adaptation in Coastal Areas, pages 337–357. Springer, 2015.

G. Borrini-Feyerabend, M.T. Farvar, J.C. Nguinguiri, and V.A. Ndangang. *Co-management of natural resources: organizing, negotiating and learning-by-doing*. Kasparek Verlag, 2000. ISBN 3-925064-30-3.

J.P. Brans and B. Mareschal. PROMETHEE Methods. In J. Figuera, S. Greco, M. Ehrgott and C. Henggeler-Antunes, editor, *Multiple Criteria Decision Analysis: State of the Art Surveys.*, pages 164–195. Springer Verlag, 2005.

J.P. Brans and P. Vincke. A Preference Ranking Organisation Method. *Management Science*, 31(6): 647–656, 1985.

C. Breidert. *Estimation of Willingness-to-Pay. Theory, Measurement and Application*. Dissertation, University Wien, 2005.

P.B. Burbridge. A generic framework for measuring success in integrated coastal management. *Ocean & Coastal Management*, 37(2):175–189, 1997.

A. Cendrero and D.W. Fischer. A procedure for assessing the environmental quality of coastal areas for planning and management. *Journal of Coastal Research*, 13(3):732–744, 1997.

F.J. Convery. Making a Difference – How Environmental Economists can Influence the Policy Process – A Case Study of David W. Pearce. *Environ. Resource Econ.*, 37:7–32, 2007.

S. Corrente, J. Rui Figueira, and S. Greco. The smaa-promethee methods. *ArXiv e-prints*, Feb 2013.

[Department for Communities and Local Government] DCLG. Multi-criteria analysis: a manual. Technical Report, DCLG, 2009.

P. Delbridge, A.J. Gray, J. Kersey, D. Luscombe, and J. Petts. Evaluating methods for public participation. R&D Technical Report E2-030/TR, 2002.

D. Drouet, J.-P. Detolle, and M. Sachs. Comparative study of information and public participation techniques in a water management context. Quebec, The Netherlands and Denmark. MEDD No BC 369, 2002.

[Deutscher Verband für Wasserwirtschaft und Kulturbau e.V.] DVWK. Ökono-mische Bewertung von Hochwasserschutzwirkungen – Arbeitsmaterial zum methodischen Vorgehen. *DVWK Mitteilungen*, 10, 1985.

A. Endres and K. Holm-Müller. *Die Bewertung von Umweltschäden*. Stuttgart, 1998.

J. Esser. Vollständigkeit, Konsistenz und Kompatibilität von Präferenzrelationen. *OR Spektrum*, 23: 182–201, 2001.

A. S. Fernndez-Castro and M. Jimnez. Promethee: An extension through fuzzy mathematical programming. *The Journal of the Operational Research Society*, 56(1):119–122, 2005.

E. Georgopoulou, Y. Sarafidis, and D. Diakoulaki. Design and implementation of a group DSS for sustaining renewable energies exploitation. *European Journal for Operational Research*, 109: 483–500, 1998.

[Joint Group of Experts on the Scientific Aspects of Marine Environmental Protection] GESAMP. The Contributions of Science to Integrated Coastal Management. Reports and Studies 61, IMO, FAO, UNESCO-IOC, WMO, WHO, IAEA, UNEP, 1996.

J.B. Grau, J.M. Anton, A.M. Tarquis, F. Colombo, L. de los Rios, and J.M. Cisneros. An application of mathematical models to select the optimal alternative for an integral plan to desertification and erosion control (Chaco Area - Salta Province - Argentina). *Biogeosciences*, 7:3421–3433, 2010.

A. Guitouni and J.-M. Martel. Tentative guidelines to help choosing an appropriate MCDA method. *Europ. J. of Operational Research*, 109:501–521, 1998.

R.M. Guzman and C.D. Kolstad. Researching Preferences, Valuation and Hypothetical Biases. *Environ. Resource Econ.*, 37:465–487, 2007.

U. Hampicke. Neoklassik und Zeitpräferenz: der Diskontierungsnebel. In F. Beckenbach, editor, *Die ökologische Herausforderung für die ökonomische Theorie*, pages 127–150. Marburg, 1991.

U. Hampicke. *Ökologische Ökonomie. Individuum und Natur in der Neoklassik*. Opladen, 1992.

B. Hansjürgens. Economic Valuation through Cost-Benefit Analysis – Possibilities and Limitations. *Toxicology*, 205(3):241–252, 2004.

V. Hartje, I. Meyer, and J. Meyerhoff. *Klimafolgen für Mensch und Küste: Am Beispiel der Nordseeinsel Sylt*, chapter Kosten einer möglichen Klimaänderung auf Sylt, pages 181–218. Springer, 2002.

M. Hisschemöller, R.S.J. Tol, and P. Vellinga. The relevance of participatory approaches in integrated environmental assessment. *Integrated Assessment*, 2(2):57–72, 2001. doi: 10.1016/j.landusepol.2004.08.011.

I.B. Hunag, J. Keisler, and I. Linkov. Multi-criteria decision analysis in environmental sciences: Ten years of applications and trends. *Science of the Total Environment*, 409(19):3578–36594, 2011.

T. Hunjak. Mathematical foundations of the methods for multi criteria decision making. *Mathematical Communications*, 2:161–169, 1997.

J. Klaus. Vorteilsanalyse zum Fachplan Küstenschutz Sylt. Technical Report, Ministry of Rural Areas, Agriculture, Food and Tourism Schleswig-Holstein, 1986.

H. Klug, M. Hamann, S. Reese, and T. Rohr. Wertermittlung für die potentiell sturmflutgefärdeten Gebiete an den Küsten Schleswig- Holsteins. Technical Report for Ministry of Rural Areas, Agriculture, Food and Tourism Schleswig-Holstein, 1998.

[Länderarbeitsgemeinschaft Wasser] LAWA. Grundzüge der Nutzen-Kosten-Untersuchungen. Technical Report, 1981.

K. Lewin and D. Cartwright. *Field theory in social science*. Harper Torchbooks, 1951.

W. Linder and A. Vatter. *Kriterien zur Evaluation von Partizipationsverfahren*, pages 181–188. Bauverlag GmbH, 1996.

J. Malczewski and C. Rinner. *Mulitcriteria Decision Analysis in Geographic Information Science*. Springer - Berlin Heidelberg, 2015. ISBN 978-3-540-74757-4. doi: 10.1007/978-3-540-74757-4.

O. Marinoni. A stochastic spatial decision support system based on PROMETHEE. *Intern. J. of Geographical Information Science*, 19(1):51–68, 2005.

D.M. Martin, V. Hermoso, F. Pantus, J. Olley, S. Linke, and L. Poff. A proposed framework to systematically design and objectively evaluate non-dominated restoration tradeoffs for watershed planning and management. *Ecological Economics*, 127:146–155, 2016. doi: 10.1016/j.ecolecon.2016.04.007.

F. Messner, O. Zwirner, and M. Karkuschke. Participation in Multicriteria Decision Support - the Case of conflicting Water Allocation in the Spree River Basin. Technical Report - Discussion Paper 6/2003 6, 2003.

Numerous Authors. Resolving environmental conflicts: Combining participation and multi-criteria analysis. In *Land Use Policy*, volume 23. Elsevier, 2006.

OECD. Citizens as Partners. OECD Handbook on Information, Consultation and Public Participation in Policy-Making. Technical report, 2001.

S.B. Olsen. Framework and indicators for assessing progress in Integrated Coastal Zone Management initiatives. *Ocean and Coastal Management*, 46:347–361, 2003.

S.B. Olsen, K. Lowry, and J. Tobey. A manual for assessing progress in coastal management. Coastal Management Report 2211, University of Rhode Island, 1999.

M.T. Ortuno. *Multi-criteria Decision Analysis (MCDA)*, pages 1376–1376. Springer Netherlands, 2013. doi: 10.1007/978-1-4020-8265-8-200870.

D.W. Pearce and R.K. Turner. *Economics of natural resources and the environment*. Harvester Wheatsheaf, 1990.

J. Petts and B. Leach. Evaluating methods for public participation: Literature Review. R&D Technical Report E135, 2000.

B. Rebolledo, A. Gil, X. Flotats, and J.A. Sanchez. Assessment of groundwater vulnerability to nitrates from agricultural sources using a gis-compatible logic multi criteria model. *Journal of Environmental Management*, 171:70–80, 2016. doi: 10.1016/j.jenvman.2016.01.041.

S. Reese, H.-J. Markau, and H. Sterr. MERK – Mikroskalige Evaluation der Risiken in überflutungsgefährdeten Küstenniederungen. Technical Report, Universiy Kiel, FTZ Westküste, 2002.

O. Renn. A model for an analytic-deliberative process in risk management. *Environmental Science and Technology*, 33(18):3049–3055, 1999. doi: 10.1021/es981283.

O. Renn. Participatory process for designing environmental policies. *Land Use Policy*, 23(1):34–43, 2006. doi: 10.1016/j.landusepol.2004.08.005.

O. Renn and A. Klinke. A framework of adaptive risk governance for urban planning. *Sustainability*, 5(5):2036–2059, 2013. doi: 10.3390/su5052036.

O. Renn and T. Webler. Konfliktbewältigung durch Kooperation in der Umweltpolitik - Theoretische Grundlagen und Handlungsvorschläge. In *Umweltökonomischen Studenteniniative OIKOS an der Hochschule St. Gallen*, pages 11–52. Ruegger Verlag, 1994.

B. Roy. Methodologie multicritere d'aide a la decision. *Economica*, XXII:423, 1985.

B. Roy. *Multi criteria Methodology for Decision Aiding*. Springer Verlag, 1996.

A. Ruhland. *Entscheidungsunterstützung zur Auswahl von Verfahren der Trinkwasseraufbereitung an den Beispielen Arsenentfernung und zentrale Enthärtung*. Dissertation, University Berlin, 2004.

E.C.M. Ruijgrok. Valuation of Nature, Water and Soil in Socio-Economic Cost Benefit. A Supplement to the Dutch Guideline for CBA. Technical Report, 2005.

P.A. Sabatier. The advocacy coalition framework: revisions and relevance for europe. *Journal of European Public Policy*, 5(1):98–130, 1998.

P. Salminen, J. Hokkanen, and R. Lahdelma. Comparing multi criteria methods in the context of environmental problems. *Europ. J. of Operational Research*, 104:485–496, 1998.

J. Salter, J. Robinson, and A. Wiek. Participatory methods of integrated assessment – a review. *WIREs Climate Change*, 1(5):697–717, 2010. doi: 10.1002/wcc.73.

J. Thomas. Force field analysis: a new way to evaluate your strategy. *Long Range Planning*, 18(6): 54–59, 1985. doi: 10.1016/0024-6301(85)90064-0.

K.D. Tucker. Force field analysis, new tool for problem solving. *Public Relations Journal*, 35(7):23, 1979.

B.A. van Asselt Marjolein and N. Rijkens-Klomp. A look in the mirror: reflection on participation in integrated assessment from a methodological perspective. *Global Environmental Change*, 12 (3):167–184, 2002. doi: 10.1016/S0959-3780(02)00012-2.

P. Walker. Dinosaur DAD and Enlightened EDD – engaging people earlier is better. *The Environmentalist*, 71:12–13, 2009.

[German Advisory Council on Global Change] WBGU. *World in Transition – Strategies for Managing Global Environmental Risks. Executive Summary*. WBGU, 1999. ISBN 3-9806309-3-5.

M. Weber. *Grundriss der Sozialökonomik. III. Abteilung: Wirtschaft und Gesellschaft*. Verlag Mohr, Tübingen, 1922.

V. Wenzel. Integrated Assessment and Multicriteria Analysis. *Phys. Chem. Earth (B)*, 26(7-8):541–545, 2001.

H. Wittmer, F. Rauschmeyer, and B. Klauer. How to select instruments for the resolution of environmental conflicts? *Land Use Policy*, 23(1):1–9, 2006. doi: 10.1016/j.landusepol.2004.09.003.

T. Würtenberger. Konfliktlösung durch Akzeptanzmanagement. In H. Zilleßen and P. Dienel and W. Strubelt, editor, *Die Modernisierung der Demokratie*, pages 72–86. Westdeutscher Verlag, 1993.

Exercises: Fictitious Case Studies

8

8.1 Introduction

These exercises could either be executed by one person or as role play by a group, if a player takes over the "role" of a representative of an organization. If these exercises are played as role play try to develop a group decision model, see, for example, Chap. 7.

All exercises are described in the same structure. It starts with a general overview of the situation and a concise description of the problem. The following paragraphs will describe rudimentary the intentions and interests of different types of land use. It is not intended to provide a comprehensive description of the different interests and needs, which would be beyond the scope of this book. On the other hand, these exercises should be open for interpretation of the respective sectors because the interests and needs of the different sectors depend strongly on the respective surrounding and circumstances in respective countries.

In general, almost every Member State (or Nation) developed a government and/or administrative system. These systems might be similar to each other or not. To apply and to execute these exercises in different governmental and administrative environments try to concentrate on the relevant issues of a problem, thus, feel encouraged to adopt the situation to specific surroundings.

Furthermore, this chapter is concentrated on the most relevant types of land use for these exercises. One can think of more interest groups which may have a stake. However, these exercises should provide a first glance of conflicts which might occur in the range of coastal management. Of course, everybody is free to reduce or extend these exercises and to adapt them to its respective intention of the class or seminar.

The following problems are modified examples from real-world problems. You might gain the feeling that some information is missing. Either additional assumptions on the

F. Ahlhorn, *Integrated Coastal Zone Management*,
Wasser: Ökologie und Bewirtschaftung, DOI 10.1007/978-3-658-17052-3_8

gaps of information have to be made or an agreement of neglecting this information could be done. At the end it is not important if you got the full information for these exercise. The main intention is to think in different roles and to try to understand different points of view on the given problem. Considering the entrance of this book where the German philosopher Kant already helped to understand that space cannot entirely be grasped or described as "thing itself", only by sensual perception and experiences.

Another point is to plan and develop an adequate strategy on how to approach the problem. What are the necessary steps? What is really important to know (relevant)? What if some steps of the process for solving the problem didn't work as expected? What if milestones or interim results won't be achieved?

8.1.1 Setting the Scene

A new institution has been installed to bridge the gap between sectoral planning and administration approaches. The institution is called *Coast*al *In*tegration *A*ssociation (CoastInA) and the founders are all relevant parties in the respective coastal zone. You are one of the first employees at the CoastInA of the southern North Sea Region and your first task is to solve two long-lasting but now getting urgent problems. The third problem refers to a situation which has been solved, maybe you will find another solution or agreement. It is important to know that we assume CoastInA has no legal sovereignty, but is acknowledged by a binding treaty of all founders, that is, CoastInA is for the time being accepted.

8.2 Problem 1: Eroding Coastal Stretch

8.2.1 Delineation of the Problem

At a coastal strip the dike fore land is eroding and half of the salt marsh has been destroyed. The stability of the main dike is at risk! In front of the dike there is a highly protected nature area. Behind the dike different types of land uses can be found: agricultural land use (e.g., meadows, dairy farming, crops) and tourism. Of course, there are more types of land use existing, but avoid to make it too complex (see Fig. 8.1). The following paragraphs describe the situation and the interests and needs of each user perspective representing different types of land use assigned to specific sectors.

8.2.2 Coastal Protection

Currently, the strategy for protecting people in the low-lying part of the coast against flooding is based on the improvement (heightening and strengthening) of the main dike. In addition to these improvement works also maintenance work in front of the dike is carried out, for example, protecting salt marshes against erosion by brush-wood groins or other

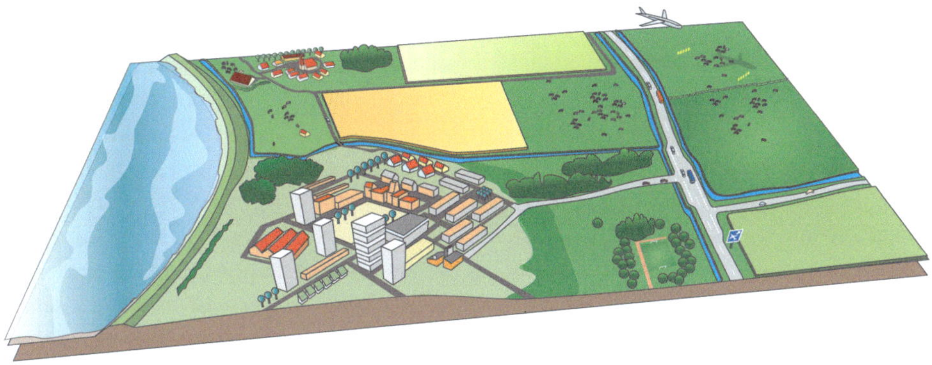

Fig. 8.1 Fictitious sketch of a coastal strip in the southern North Sea. Developed within the framework of the EU Interreg IIIB project ComCoast. ©van Lint Vormgeving 2006

nature-oriented measurements. At the specific spot you are dealing with the envisaged heightening of a dike to the required safety standard including a safety margin for an accelerated sea level rise (SLR) of about 50 cm. Besides the national rules (which may differ between the Member States) the European Flood Risk Management Directive (see Sect. 6.1.3) has to be implemented and considered.

Organizations involved for the sector Coastal Defense (Protection) are the dike board, the responsible technical administrative body and the ministry of water and environment.

8.2.3 Agricultural Use

Parts of the area behind the dike are intensively used to grow crops and for the cultivation of energy plants. In the proximity there is a biomass power plant producing energy as a joint venture of local farmers. The area was gained by land reclamation works centuries ago and, therefore, consists of highly fertile soil. Besides crop land a traditional dairy farmer can be found. He has recently invested in a new milking robot and bought the land from older farmers who were retired.

Organizations involved for the sector agriculture are the farmer association for the biomass power plant, the chamber of agriculture (as the technical advice body for farmers and the respective ministry) and the peasantry of that area.

8.2.4 Nature Conservation

In front of the dike a national park has been established 30 years ago. The status of the protection is divided into three categories: high, medium and open for leisure activities. A few years ago the area of the national park has been designated as World Heritage area. Additionally, since 20 years a biosphere reserve covers the main area of the national park which radiates to adjacent areas.

Behind the dike a complex network of nature and landscape conservation areas were developed over time. These areas show overlap with compensation measurements for several infrastructure and development projects. Besides the national regulations the consideration of the European Water Framework Directive (see 6.1.1) and its daughter Directives are mandatory.

Different organizations are representing this sector, for example, NGOs (local or national), nature conservation administration, a technical advice body for the government and district government as lowest part of the different administration levels. Additionally, the Ministry of Water and Environment has to be involved. Furthermore, nature conservation is a crucial part of the spatial planning system, thus, the different levels responsible for spatial/physical planning should be involved, too.

8.2.5 Water Management

The description of the water management sector is split into two parts: (i) drainage management and (ii) ground water management (i.e., drinking water supply).

Drainage Management

In the area of concern the surface water management is conducted by self-organized bodies since centuries. The drainage system grew about the last 100 years to an efficient system of ditches and canals, pumping stations and locks. The basic philosophy of the responsible water board is to drain the surface water as fast as possible out of both rural and urban area. With the implementation of the Water Framework Directive (see 6.1.1) and some new regulations from physical planning, for example, to build retention ponds in new housing or industrial areas, the original system works mainly to fulfill the aims of the water boards. One of the main tidal gates to drain the area of the water board is in the dike stretch of concern.

Organizations to be considered from drainage management are the water board, a technical advisory body for the administration and the ministry of water and environment.

Ground Water Management

In ancient times people at the coastal strip gained their drinking water out of cisterns and small retention ponds because the water in the ditches was saline. Over time the population grew and the provision of the people with drinking water was limited. A company for drinking water was founded and installed several wells in the coastal area to extract drinking water from the ground water body. One of the drinking water works with several wells is in your area of concern.

Organizations you have to deal with are the company for drinking water supply and the district government which is responsible for the permission of extracting ground water and the security of public supply.

8.2.6 Tourism

Although the area of concern is dominated by agricultural use the tourism manager wants to improve the offers and opportunities for recreation. Some places have already been identified to be developed for recreational purposes. It is yet not finally decided, but specific groups are concerned by the new developments. The main attraction point is the advertisement of the designated World Heritage area. The entire coastal zone (including the islands in front of the main land) has been roughly divided into different *hotspots* for recreational purposes, such as nature-oriented recreation, bed-and-breakfast or emphasizing on the traditional landscape and living.

The organizations which have to be considered are scattered because each municipality maintains a tourist office, the district council is responsible for spatial planning and, thus, for recreation and tourism. Some districts established a tourism association but without any sovereignty, only for the advertisement of the coastal stretch. Furthermore, after the designation as World Heritage area also the national park administration and an international office for the coordination of cross-border ambitions in nature conservation are developing touristic strategies.

8.2.7 Tasks

1. Write down what you want to achieve
2. Develop a plan on how to start tackling the existing situation
3. Draw a sketch of the process you want to apply
4. Make an overview of what has to be considered linked to both regulations and interests and needs of sectors affected

8.3 Problem 2: Small Harbor Accessibility

8.3.1 Delineation of the Problem

As experienced employee of the well-known and acknowledged CoastInA you have been approached by a small municipality to help solving a problem they are facing. In a meeting the mayor explained the situation as follows:

The municipality is lying directly at the coast. The main type of land use is farming, mainly dairy farming. But also the growth of energy plants is increasing. Some small villages are located in the municipality, but no city or capital. The area has some touristic features but not so many. One of the main attraction points is a small harbor for fishing and some leisure boats. The fishermen use cutters to catch different sorts of fish in the adjacent water bodies of the North Sea. It is estimated that two-thirds of the total number of tourists are visiting this small harbor at least once during their stay.

The access of the small harbor via a tidal channel which is part of the mouth of an estuary discharging into the North Sea (Fig. 8.2). Since decades the access to the small harbor is getting worse due to siltation. The situation of the catchment area of this tidal channel has dramatically changed over the last decades due to several changes in the mouth of the estuary. One of the major changes has been the straightening and deepening of the main channel of the river. In ancient times, this channel has been decided to be the fairway for sailing boats. Currently, it is used for container vessels and bulk carrier to the big harbors in the estuary. Consequently, the side channels, and also the tidal channel to the small harbor, suffered higher sedimentation rates, for example, due to a decreased current flow velocity.

Currently, the municipality fears that the situation in the tidal channel will get even worser and in consequence the access to their small harbor will be interrupted. One of the

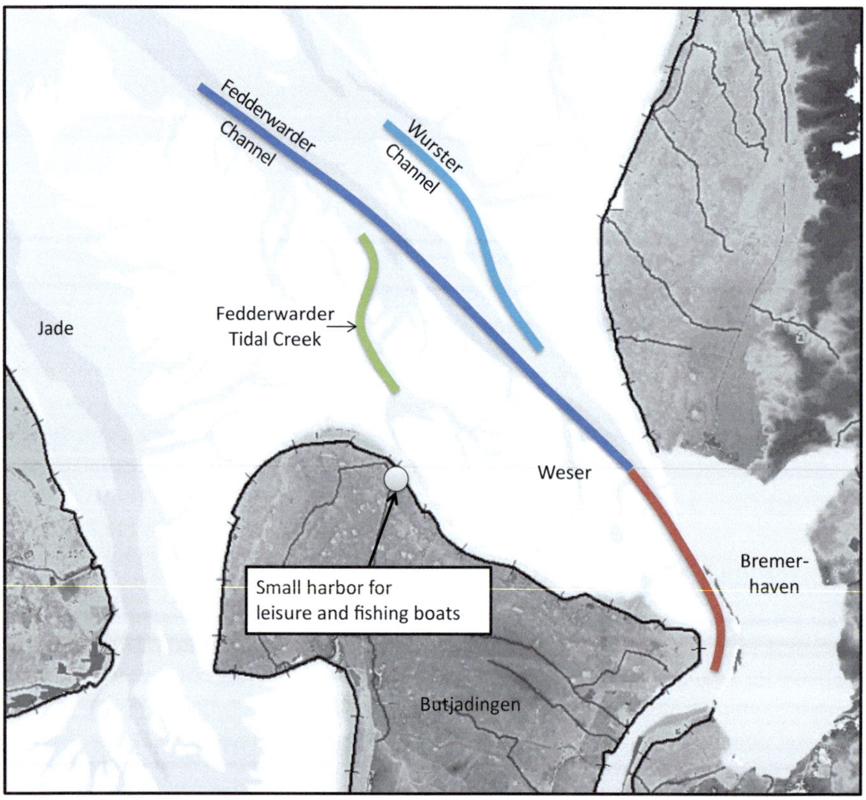

Fig. 8.2 Channel system of the outer Weser excerpt from a map compiled by Ahlhorn et al. (2010). There are two channels, the "Wurster Channel" and the "Fedderwarder Channel", where the latter is the main fairway for container ships and vessels to the harbors Bremerhaven and Bremen down the river Weser. The southern channel, called Fedderwarder Priel (tidal creek), is the divergence to the small harbor at the peninsula of Butjadingen.

expected consequences is the emigration of the fishermen and the cutters, and thus, of an the loss of the important attraction.

8.3.2 Municipality

The municipality is not owner of the harbor infrastructure. But it is responsible for the physical planning and the development of a touristic strategy with their tourism company.

Due to the interlinkage with the fairway of a river different administrative levels are concerned. The national level, because the fairway of a river is under jurisdiction of the national government. The Federal State or provincial level, because the area adjacent to the fairway is under responsibility of the federal state or provincial government.

8.3.3 National Level: Shipping Administration

In your case a Water and Shipping Administration is the responsible national body for deepening, straightening and maintenance of the fairway. The reason for these river works is given by the application of the Federal State Ministry for Economic Affairs and Transport which aims at providing the best circumstances for the harbors along the river. The Water and Shipping Administration gets scientific support for all tasks linked to implementation of river works by a national research institute. They are, for example, conducting morphological studies and running and operating different simulation models. The planning and implementation has to be done within the framework of the WFD (see Sect. 6.1.1) and other national and European regulations.

8.3.4 Federal State/Provincial Level: Water/Environment

There is a technical agency supporting the local, regional and Federal State administration with their expertise on, for example, morphological development in rivers and shallow sea water, coastal defense and water management in general. Furthermore, it has also to be involved concerning the questions of WFD (i.e., chemical and ecological quality) outside the fairway in the river.

8.3.5 Federal State/Provincial Level: Harbor Authority

The infrastructure of this small harbor and a small strip of 200 m around the harbor belongs to the Ministry of Finance. The Federal State harbor authority is responsible for the administration on behalf of this Ministry. The supervision of the harbor authority is done by the Ministry of Economic Affairs and Transport.

8.3.6 Drainage Management

The local water board uses the tidal inlet of the harbor for draining purposes. in the adjacent dike there is a tidal gate which opens passively during low tide and closes with the rising water level. The inland water flushes the short tidal inlet of the harbor. An agreement exists between the water board and the harbor authority to drain as much inland water through the tidal gate as possible.

8.3.7 Tasks

1. Draw a sketch about the first steps to be taken.
2. Prepare a flow chart for the working process.
3. Write down the different argumentation lines why the different administrative bodies might be reluctant to take the full responsibility for the current situation of the side channel which is the access to the small harbor. How to overcome them?

8.4 Problem 3: Closure of a Bay

Due to the storm surges in the Middle Ages the area of many bays along the southern North Sea coast was enlarged. Here, the exercise deals with the closure of the Ley Bay (Fig. 8.3) which is part of East Frisia in north-western Lower Saxony. The description of the exercise is taken from Ahlhorn (2009, p. 14):

> The Ley Bay witnessed its largest extension approx. 600 years ago, as a consequence of severe storm tides in the middle ages. Until the middle of the last century land reclamation works were executed to increase the arable area for the inhabitants. In the 1950s the Ley Bay was mainly shaped by economic drivers (Erchinger, 1970; Hartung, 1983; Janssen, 1992; Kunz, 1999). Approx. 10,000 ha were reclaimed and this new land was offered to inhabitants and refugees of World War II. After the 1950s the effort to reclaim land from the sea decreased, because of the diversification of working fields after World War II and the increase of the effectiveness in agriculture. On the other hand, the problems with water management in the hinterland around the Ley Bay intensified, the existing tidal channels silted up continuously. Until 1985 the problems also affected the harbors around the Ley Bay which were dependent on free access to the North Sea. Siltation had imposed increasing pressure on water management and shipping, resulting in the installation of pumping stations with continuously increasing performance. (Janssen, 1992)
>
> In the 1960s a plan was developed to enclose the Ley Bay, because of sedimentation problems: limited access of ships to the harbors and limitation of the drainage of the hinterland. The Master Plan for Coastal Protection of 1973 stated that the height of the main dike was not sufficient (Nds. MELF, 1973). These enclosure plans were also mentioned in regional planning programs. The first idea for the enclosure of the Ley Bay took into account different types of

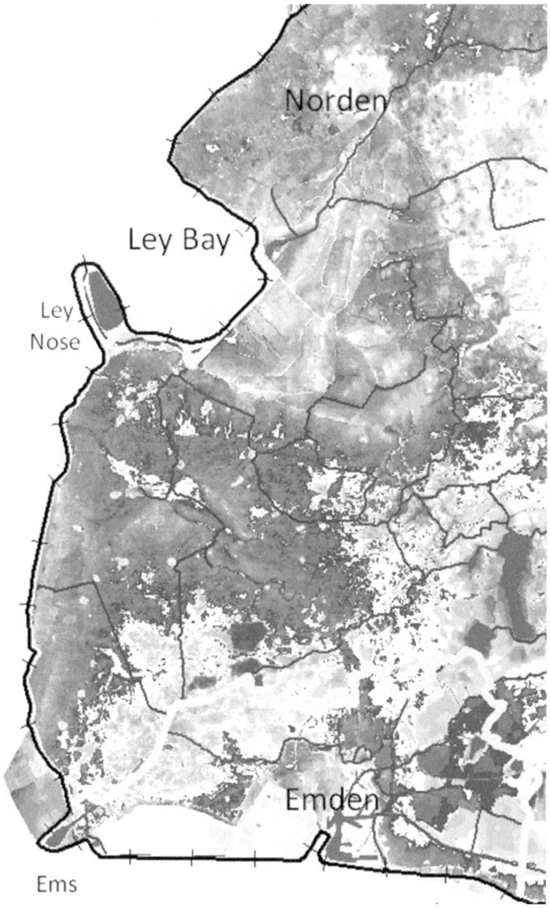

Fig. 8.3 The Ley Bay in Lower Saxony (Germany) is located north of the river Ems between the cities of Emden and Norden. Source: adapted from Ahlhorn et al. (2010)

land use, such as agriculture, recreation and nature conservation. But, the survey only took the development of the enclosed bay into account, i.e. after a certain time the ecosystem would have changed from a saline to a freshwater environment. This would have caused a tremendous change of the flora and fauna of the existing ecosystem. In the late 1970s the interests of nature conservation, especially for the conservation of the Wadden Sea habitats, attracted increased attention. The establishment of these interests demanded new solutions to the Ley Bay problems, like water management and shipping. In 1980 the government of Lower Saxony canceled the planning approval of a complete enclosure and recommended to maintain the natural status of the Ley Bay as saline environment. Afterwards, a group consisting of engineers as well as representatives from nature conservation bodies agreed on a compromise for the required work in the Ley Bay, the "Ley nose" (Fig. 8.4). The issue was taken to the European Court of Justice in Luxembourg which decided that only outstanding reasons can lead to the reduction of conservation areas, whereas economic aspects are irrelevant. This decision has had

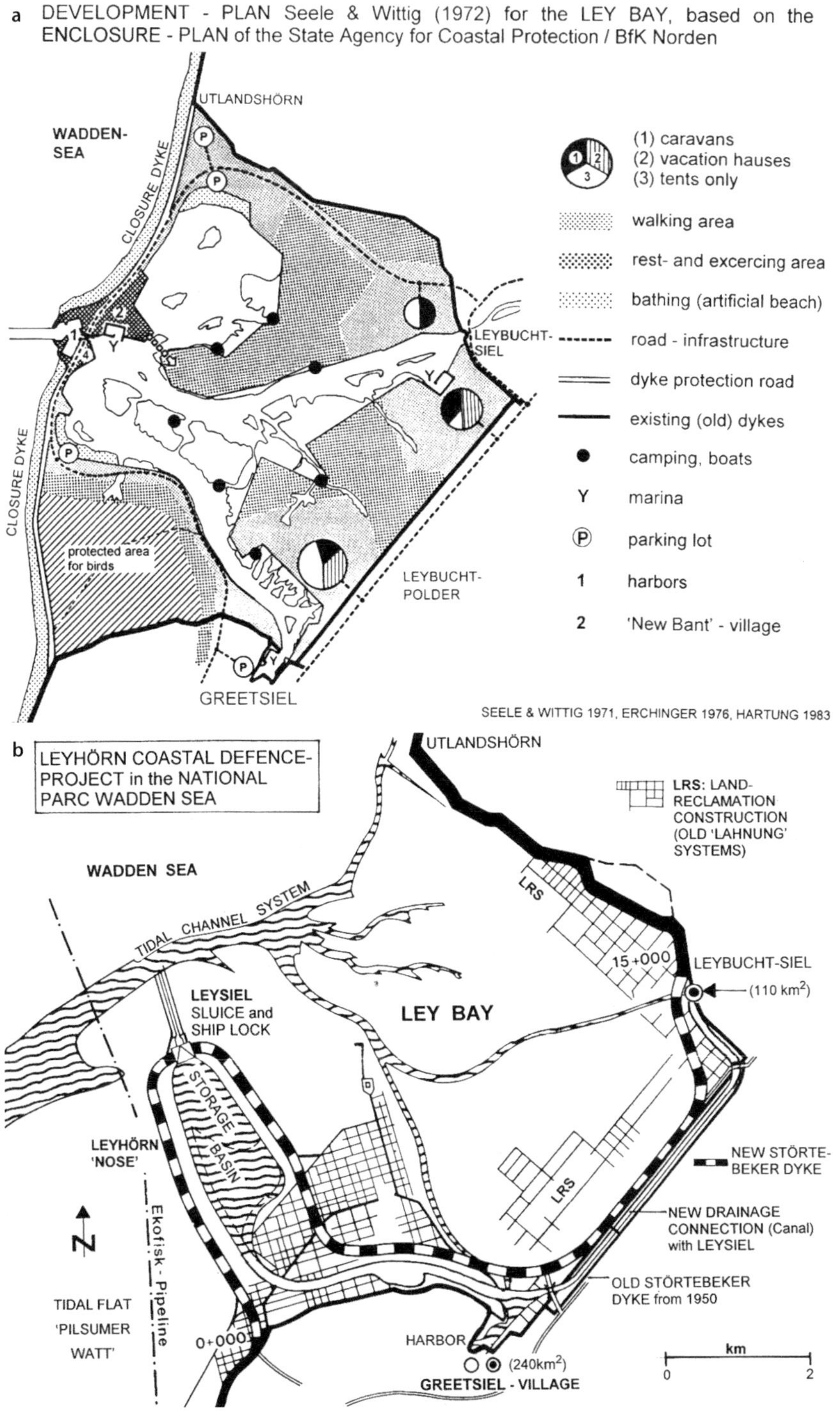

Fig. 8.4 Solution approaches for the Ley Bay: (a) The first proposed solution as closure of the entire bay with multifunctional use (*top*), (b) the final compromise after approx. 15 years of negotiations (*bottom*). Source: Kunz (1999)

large-scale consequences for further (coastal protection) projects in the entire European Union. Today, the Ley Bay project is completed and the Ley nose has already been built – Janssen (1992). (Ahlhorn (2009, p. 160–161)

8.4.1 Task

In Fig. 8.4 the first proposal and the finally implemented solution are shown. Compare both approaches.

1. What might be the reasons why the first proposal as multifunctional approach (A) has been rejected by representatives of environmental groups?
2. What might be beneficial aspects of the compromise (B)?

References

F. Ahlhorn. *Long-term Perspective in Coastal Zone Development. Multifunctional Coastal Protection Zones*. Springer, 2009.

F. Ahlhorn, J. Meyerdirks, and I. Umlauf. Speichern statt pumpen. Abschlussbericht. Technical report, University of Oldenburg, 2010.

H.-F. Erchinger. Küstenschutz durch Vorlandgewinnung, Deichbau und Deicherhaltung in Ostfriesland. *Die Küste*, 19:125–185, 1970.

W. Hartung. Die Leybucht (Ostfriesland) – Probleme ihrer Erhaltung als Naturschutzgebiet. *Neues Archiv für Niedersachsen*, 32(4):355–387, 1983.

T. Janssen. *Die Leybucht: Natur- und Küstenschutz. Eine Dokumentation*. SKN, Norden, 1992.

H. Kunz. The Leyhörn Project – An integrated engineering response to multifarious demands of a coastal community. In *Proc. Fifth Intern. Conf. on Coastal and Port Engineering in Developing Countries*, volume 3, pages 2176–2186. COPEDEC, 1999.

[Niedersächsisches Ministerium für Ernährung, Landwirtschaft und Forsten] Nds. MELF. Generalplan Küstenschutz Niedersachsen. Technical report, Niedersächsisches Ministerium für Ernährung, Landwirtschaft und Forsten, 1973.

Index

F. Ahlhorn, *Integrated Coastal Zone Management*,
Wasser: Ökologie und Bewirtschaftung, DOI 10.1007/978-3-658-17052-3

MIX
Papier aus verantwortungsvollen Quellen
Paper from responsible sources
FSC® C105338

FSC
www.fsc.org

If you have any concerns about our products,
you can contact us on
ProductSafety@springernature.com

In case Publisher is established outside the EU,
the EU authorized representative is:
Springer Nature Customer Service Center GmbH
Europaplatz 3, 69115 Heidelberg, Germany

Printed by Libri Plureos GmbH
in Hamburg, Germany